U0173593

高等学校城乡规划专业系列推荐教材

城乡规划专业英语

阎传海　于洪蕾　田　华　朱一荣　编著

中国建筑工业出版社

图书在版编目（CIP）数据

城乡规划专业英语 / 阎传海等编著．— 北京：中国建筑工业出版社，2022.4
高等学校城乡规划专业系列推荐教材
ISBN 978-7-112-27294-5

Ⅰ．①城…　Ⅱ．①阎…　Ⅲ．①城乡规划—英语—高等学校—教材　Ⅳ．① TU98

中国版本图书馆 CIP 数据核字（2022）第 060652 号

本书包括上、下两篇。上篇为核心词汇分类。城乡规划专业英语核心词汇是指城乡规划专业英语词汇中的词根词与前缀词。基于 GRE、GMAT、SAT、ACT、TOEFL、IELTS 的核心词汇构成，结合城乡规划经典文献词汇特点，本书确定城乡规划专业英语核心词汇包括 78 个前缀，235 个词根。词根词与前缀词具有显著的相似性，可举一反三，可进行推理，能显著地提高记忆速度。下篇为经典文献导读。本书根据城市规划理论发展演变与城市规划学科发展脉络选择阅读文献，划分成 20 个单元，每个单元包括课文、注释、生词（课文中黑体字注出生词）、复杂长句解析（课文中下划线注出复杂长句，本书共鉴别、解析 69 句复杂长句）以及习题。

本书的特点与目的：创新核心词汇的记忆模式，解析复杂长句的结构奥秘，提升专业文献的阅读能力，把握城乡规划的发展脉络。

本书可作为城乡规划、人文地理本科生、研究生的教材，可供建筑学、风景园林本科生、研究生以及相关专业人员参考使用。

为更好地支持本课程的教学，我们向使用本书的教师免费提供教学课件，有需要者请与出版社联系，邮箱：jgcabpbeijing@163.com。

责任编辑：杨　虹　尤凯曦　牟琳琳
责任校对：王　烨

高等学校城乡规划专业系列推荐教材
城乡规划专业英语

阎传海　于洪蕾　田　华　朱一荣　编著
*
中国建筑工业出版社出版、发行（北京海淀三里河路 9 号）
各地新华书店、建筑书店经销
北京雅盈中佳图文设计公司制版
河北鹏润印刷有限公司印刷
*
开本：787 毫米 × 1092 毫米　1/16　印张：$23^1/_4$　字数：473 千字
2022 年 4 月第一版　2022 年 4 月第一次印刷
定价：**58.00** 元（赠教师课件）
ISBN 978-7-112-27294-5
（39105）

代序

核心词汇、复杂长句与经典文献

一、核心词汇

本书的设计、完成源于我们2011年初开始的GRE词汇研究。

GRE，Graduate Record Examination，即通常所说的美国研究生入学考试。GRE词汇为学术英语词汇，对于专业英语的学习具有重要的意义。

GRE词汇量大，单词结构复杂、生僻、语义抽象。《GRE词汇精选》（俞敏洪，2008）共580页，收录单词7463个（仅指每页左侧所列单词，下同）。《NEW GRE词汇精选》（俞敏洪，2011）共470页，收录单词6524个（删除2254个，新增1315个）。两本书共收录8778个单词。学习GRE词汇是一项极其艰巨的任务。

根据多年生态学研究的思路与方法，我们对GRE词汇的组成结构、词间关系进行了研究，将GRE词汇系统地划分为词根词、前缀词、同形词、学科词、同义词、相关词等（《GRE词汇分类记忆宝典》，阎传海，2013）。每类单词，特别是词根词、前缀词类型的单词，具有显著的相似性，可举一反三，可进行推理，能显著地提高记忆速度。我们称此种单词记忆方法为类群记忆法，即单词的记忆可一类一类地记，一群一群地记。类群记忆法为每类单词设计了一个"记忆核"，单词的记忆犹如在"记忆核"的基础上滚雪球，这是类群记忆法之所以能显著提高记忆速度的原因。类群记忆法属于理解性、推理性英语单词记忆方法。现以词根"gress"与"vor"为例说明如下（解析式中，后面有短横"-"的，如ag-，di-，ex-，in-，trans-，omni-等，表示前缀；后面没有短横"-"的，如gress，carn，herb，vor等，表示词根或单词）。

87. gress 走

aggression	n. 侵略；敌对的情绪或行为【ag-（加强语气）+ gress（走）+ ion →走到别人的地盘→侵略】
aggressive	adj. 好斗的；进取的
aggressively	adv. 侵略地，攻击地
aggressor	n. 侵略者，攻击者
digress	v. 离题【di-（二）+ gress（走）→离题】
digression	n. 离题，题外话
digressive	adj. 离题的，枝节的
egress	n. 出去，出口【e-（e-= ex-，出，外）+ gress（走）→出去，出口】
ingress	n. 进入【in-（在……内）+ gress（走）→进入】
regress	v. 使倒退，复原，逆行【re-（回，反）+ gress（走）→倒退】
regressive	adj. 退步的，退化的
transgress	v. 冒犯，违背【trans（超过）+ gress（走）→走过了→超越界限→冒犯，违背】
transgression	n. 违法，罪过

235. vor 吃

carnivorous	adj. 肉食动物的【carn（肉）+ i + vor（吃）+ ous →肉食动物的】
herbivorous	adj. 食草的【herb（草）+ i + vor（吃）+ ous →食草的】
omnivorous	adj. 杂食的；兴趣杂的【omni-（全部）+ vor（吃）+ ous →杂食的】
devour	v. 吞食；如饥似渴地读
voracious	adj. 狼吞虎咽的，贪婪的
voracity	n. 贪婪

快速突破 GRE 词汇的奥秘在于：熟记 612 个词根、78 个前缀，理解、读熟词根词与前缀词（在理解的基础上多读几遍）。如此，从数量而言，掌握了近 50% 的 GRE 词汇；从难度而言，可能掌握了 70%~80% 的 GRE 词汇，因为其他类型的 GRE 词汇，如学科词、同义词、相关词等，大多组成简单，含义明确。

其后，我们对 *The OFFICIAL GUIDE to the GRE revised General Test*、*BARRON'S NEW GRE*：*GRADUATE RECORD EXAMINATION*（*19TH EDITION*）等多种著作中的 GRE 词汇进行了分析，发现词根词、前缀词是其核心。因此，我们把词根词、前缀词称为核心词汇。这不仅适用于 GRE 词汇，同样适用于 GMAT 词汇、SAT 词汇、ACT 词汇、TOEFL 词汇、IELTS 词汇以及大学四、六级词汇，当然也适用于城乡规划专业英语词汇。

本书基于GRE、GMAT、SAT、ACT、TOEFL、IELTS的核心词汇构成，结合城乡规划经典文献词汇特点，确定城乡规划专业英语核心词汇包括78个前缀，235个词根（详见上篇）。

希望同学们能够背下这些前缀与词根，能够重视分析每个前缀、每个词根下所列单词与该前缀、该词根的关系，能够在理解的基础上记住这些单词。这样，你的词汇量就会有质的飞跃，专业文献阅读的词汇问题也就基本解决了。

二、复杂长句

除词汇外，复杂长句是GRE、GMAT阅读能力提升的另一个瓶颈，当然也是城乡规划专业英语阅读能力提升的另一个瓶颈。有学者专门对GRE、GMAT阅读的复杂长句进行了研究（《GRE & GMAT阅读难句教程》，杨鹏，2015）。在对GRE句子填空研究的基础上（《GRE句子填空对策与技巧》，阎传海，2016），详细研读了《GRE & GMAT阅读难句教程》，我们认为，复杂长句攻克的关键在于对其结构的把握，具体而言，就是基于英语语法规则，对复杂长句进行拆分，并明晰各拆分部分之间的关系，如此，复杂长句也就读懂了。现以5个GRE、GMAT复杂长句为例说明如下：

1. The methods that a community devises to perpetuate itself come into being to preserve aspects of the cultural legacy that that community perceives as essential.

2. Virgin-soil epidemics are those in which the populations at risk have had no previous contact with the disease that strike them and are therefore immunologically almost defenseless.

3. Perhaps the fact that many of these first studies considered only algae of a size that could be collected in a net (net phytoplankton), a practice that over-looked the smaller phytoplankton (nannoplankton) that we now know grazers are most likely to feed on, led to a de-emphasis of the role of grazers in subsequent research.

4. A long-held view of the history of the English colonies that became the United States has been that Englands's policy toward these colonies before 1763 was dictated by commercial interests and that a change to a more imperial policy, dominated by expansionist militarist objectives, generated the tensions that ultimately led to the American Revolution.

5. These questions are political in the sense that the debate over them will inevitably be less an exploration of abstract matters in a spirit of disinterested inquiry than an academic power struggle in which the careers and professional fortunes of many women scholars-only now entering the academic profession in substantial numbers-will be at stake.

解析如下：

1. **The methods** that a community devises to perpetuate itself **come into being to preserve aspects of the cultural legacy** that that community perceives as essential.

黑体字部分为句子的主干（主语、谓语和宾语），第一个 that 引导的从句为定语从句，修饰其前的 The methods，第二个 that 引导的从句亦为定语从句，修饰前面的 aspects of the cultural legacy。

2. **Virgin-soil epidemics are those** in which the populations at risk have had no previous contact with the disease that strike them **and are therefore immunologically almost defenseless.**

黑体字部分为句子的主干（主语、谓语和表语），which 引导的定语从句修饰其前的 those，that 引导的定语从句修饰前面的 the disease。

3. Perhaps **the fact** *that many of these first studies considered only algae of a size that could be collected in a net (net phytoplankton)*, a practice that over-looked the smaller phytoplankton (nannoplankton) that we now know grazers are most likely to feed on, **led to a de-emphasis of the role of grazers in subsequent research.**

句子结构非常复杂。黑体字部分为句子的主干（主语、谓语和宾语）。斜体字部分为 that 引导的定语从句，修饰其前的 the fact，其中又包含一个 that 引导的定语从句，修饰前面的 algae of a size。a practice 是主语 the fact 的同位语，其被 that 引导的定语从句所修饰，而该定语从句又包括一个 that 引导的定语从句，修饰其前的 the smaller phytoplankton (nannoplankton)。

4. **A long-held view of the history of the English colonies** that became the United States **has been** that Englands's policy toward these colonies before 1763 was dictated by commercial interests and that a change to a more imperial policy, dominated by expansionist militarist objectives, generated the tensions that ultimately led to the American Revolution.

黑体字部分为句子的主语和谓语，the English colonies 被 that 引导的定语从句所修饰，两个并列的 that 引导的从句为表语从句，第二表语从句中 dominated 引导的过去分词短语作定语，修饰其前的 a change。

5. **These questions are political** *in the sense that* ***the debate over them will inevitably be less an exploration of abstract matters in a spirit of disinterested inquiry than an academic power struggle*** *in which the careers and professional fortunes of many women scholars-only now entering the academic profession in substantial numbers-will be at stake*.

less...than... 结构，意即“与其说是……还不如说是……”。

正体黑体字部分为句子的主干（主语、谓语和表语）。斜体字部分为 in the sense that 引导的原因状语从句，整体结构为 less...than... 结构，斜体黑体字部分为原因状语从句的主干，which 引导的定语从句被两个破折号分开，破折号之间的内容为解释性插入语，定语从句修饰其前的 an academic power struggle。

句子结构非常复杂，参考译文如下：这些问题具有政治性，因为对于这些问题的争论将不可避免地成为，与其说是在一种公平探讨精神下对抽象事务的一种探索，还不如说是一场学术权力的斗争，在这场斗争中很多女性学者（只是现在才大量进入学术界）的职业命运将会处于危机之中。

读懂是目的，但不是结束。杨鹏认为，复杂长句应反复阅读，每个句子至少读到 20 遍以上，甚至背诵下来，这样你的阅读能力就会在不知不觉中得到大幅度的提升，以前看的似懂非懂的文章一下子就清晰了。我们深有体会，赞同这一观点，希望同学们身体力行。

本书共鉴别、解析 69 句复杂长句，详见下篇各单元。

三、经典文献

如何选择阅读文献是我们重点研究的问题。我们认为，选择的阅读文献应反映城市规划理论的发展演变，应体现城乡规划学科的发展脉络，使学生既能提升专业文献阅读能力，又能提高城市规划理论修养。

1898 年 Ebenezer Howard 出版 *Tomorrow: A Peaceful Path to Real Reform*，1902 年以 *Garden Cities of Tomorrow* 为书名再版，提出田园城市理论。百余年来，田园城市理论对世界城市规划理论与城市建设实践产生了广泛而深远的影响。今天的规划界一般把 Howard 田园城市理论的提出作为现代城市规划的开端。

19 世纪末至 20 世纪末西方城市规划理论的发展演变大致分为三个阶段：

第一阶段（19 世纪末至 20 世纪 60 年代末）：该阶段城市规划特征主要包括三个方面：①认为城市规划是物质环境规划（the physical planning），并因此与社会规划（the social planning）和经济规划（the economic planning）区分开来；②以设计为核心（the designed-based view of planning），视城市规划为建筑设计的扩大化；③认为城市规划工作者应该像建筑师或者工程师那样为城市土地使用和空间形态的构架提供终极状态（end-state）的总体（the master planning）规划或蓝图规划（the blueprint planning）。该阶段代表性城市规划理论包括田园城市理论、卫星城市理论（satellite towns）、新城理论（new towns）、区域规划理论（regional planning）、调查—分析—规划模式（survey-analysis-plan）、现代主义城市规划理论（modernism）[例如当代城市（the contemporary city）、伏瓦生规划（Plan Voison）、光辉城市（the radiant city）、雅典宪章（Athens Charter）]、邻里单位理论（the neighborhood unit）、雷德朋原则（the Radburn principles）、广亩城市理论（the broadacre city）等。该阶段代表性城市规划理论家包括 Howard，Unwin，Geddes，Le Corbusier，Perry，Wright，Mumford，Keeble 等。

第二阶段（20世纪60年代）：该阶段是西方城市规划理论发展的重要转折时期，在对前一阶段城市规划理论批判的基础上，城市规划由设计转变为科学，并认为城市规划是一个政治过程。该阶段代表性城市规划理论包括系统规划理论（the systems view of planning）、理性过程规划理论（the rational process view of planning）、渐进性规划理论（the incremental planning）、混合审视模式（the mixed scanning model）、倡导性规划理论（the advocacy planning）等。该阶段代表性城市规划理论家包括McLoughlin，Chadwick，Lindblom，Jacobus，Alexander，Davidoff，Etzioni，Arnstein等。

第三阶段（20世纪70年代至20世纪末）：该阶段对理性规划模式进行了批判，认为理性规划模式忽视了对城市规划是如何实施的这一至关重要问题的关注；发现规划的有效实施与人们的相互交流和协商技巧密切相关，这推动了协商规划理论（the transactive planning）或沟通规划理论（the communicative planning）的发展。此外，新马克思主义理论（the neo-Marxist approach to planning）、生态城市理论（eco-cities）、新城市主义理论（new urbanism）、可持续发展理论（sustainable development）等城市规划理论的发展也是该阶段的显著特征。该阶段代表性城市规划理论家包括Sager，Innes，Healy，Harvey，Castells，Register，Katz，Duany，Dutton，Calthorpe，Wheeler等。

重要城市规划理论家及其代表性著作（文章）见下表（详见**Bibliography and References**）：

Time	Author	Title
1870	Olmsted，F. L.	*Public Parks and the Enlargement of Towns*
1889	Sitte，C.	*City Planning According to Artistic Principles*
1898	Howard，E.	*Garden Cities of Tomorrow (original title: Tomorrow: A Peaceful Path to Real Reform)*
1909	Unwin，R.	*Town Planning in Practice: An Introduction to the Art of Designing Cities and Suburbs*
1915	Geddes，P.	*Cities in Evo1ution: An introduction to the Planning Movement and the Study of Civics*
1924	Le Corbusier	*The City of Tomorrow*
1933	Le Corbusier	*The Radiant City*
1929	Perry，C.	*The Neighborhood Unit. Neighborhood and Community Planning: Regional Survey Volume Ⅶ, Regional Plan of New York and Its Environs*
1935	Wright，F. L.	*Broadacre City: A New Community Plan*
1938	Mumford，L.	*The Culture of Cities*
1952	Keeble，L.	*Principles and Practice of Town and Country Planning*

续表

Time	Author	Title
1959	Lindblom, C. E.	*The science of "muddling through"*
1961	Jacobs, J.	*The Death and Life of Great American Cities*
1961	Mumford, L.	*The City in History: Its Origins, Its Transformations, and Its Prospects*
1961	Lynch, K. A.	*The Image of the City*
1963	Buchanan, C.D. et al.	*Traffic in Towns*
1965	Alexander, C.	*A city is not a tree*
1965	Davidoff, P.	*Advocacy and pluralism in planning*
1966	Venturi, R.	*Complexity and Contradiction in Architecture*
1967	Etzioni, A.	*Mixed-scanning: a "third" approach to decision-making*
1969	Arnstein, S. R.	*A Ladder of Citizen Participation*
1969	McLoughlin, J. B.	*Urban and Regional Planning: A Systems Approach*
1971	Chadwick, G. F.	*A Systems View of Planning*
1973	Faludi, A.	*Planning Theory*
1973	Harvey, D.	*Social Justice and the City*
1977	Castells, M.	*The Urban Question: A Marxist Approach*
1984	Lynch, K. A.	*Site Planning. 3rd edition*
1988	Lynch, K. A.	*Good City Form*
1987	Register, R.	*Eco-city Berkeley: Building Cities for a Healthier Future*
1987	Forester, J.	*Planning in the face of conflict*
1989	Forester, J.	*Planning in the Face of Power*
1994	Katz, P.	*The New Urbanism: Toward an Architecture of Community*
1994	Sager, T.	*Communicative Planning Theory*
1995	Innes, J. E.	*Planning theory's emerging paradigm: communicative action and interactive practice*
1997	Healy, P.	*Collaborative Planning: Shaping Places in Fragmented Societies*
1988	Friedmann, J.	*Planning in the Public Domain: From Knowledge to Action*
1998	Taylor, N.	*Urban Planning Theory since 1945*
2000	Duany, A. et al.	*Suburban Nation: The Rise of Sprawl and the Decline of the American Dream*
2000	Dutton, J. A.	*New American Urbanism: Re-forming the Suburban Metropolis*
2001	Calthorpe, P. et al.	*The Regional City: Planning for the End of Sprawl*
2002	Hall, P.	*Urban and Regional Planning. 4th edition*
2002	Hall, P.	*Cities of Tomorrow: An Intellectual History of Urban Planning and Design in the Twentieth Century. 3rd edition*
2004	Wheeler, S. et al.	*The Sustainable Urban Development Reader*

综上所述，本书的特点与目的可概括如下：创新核心词汇的记忆模式，解析复杂长句的结构奥秘，提升专业文献的阅读能力，把握城乡规划的发展脉络。

本书可作为城乡规划、人文地理本科生、研究生的教材，可供建筑学、风景园林本科生、研究生以及相关专业人员参考使用。

编写说明

1. 后面有短横“-”的，如 ag-，di-，ex-，in-，trans-，omni- 等，表示前缀；

2. 后面没有短横“-”的，如 gress，carn，herb，vor 等，表示词根或单词；

3. 词表中单词后“【 】”中的内容为该单词的解析式，即根据单词中包含的词根、前缀推导出该单词的含义，建议予以充分重视；

4. 每个单元（**Unit**）包括课文（**Text**）、注释（**Notes**）、生词（**New Words**）、复杂长句解析（**Complex Sentences**）与练习（**Exercises**）等；

5. 课文中生词加粗注出，复制置放于“**New Words**”下，解除加粗，鉴别词根词与前缀词，加粗词根词与前缀词，同时注出词根与前缀，与上篇相对应；

6. 课文中下划线部分为有关城市规划理论重要论述，需强化阅读，每篇课文习题部分均要求将下划线句子或段落译成中文；

7. 课文中下划线斜体字部分为复杂长句，其解析详见“**Complex Sentences**”；

8. 参考文献分上、下篇列出，注意下篇参考文献事实上包括两部分：第 [14] 至第 [24] 为一部分，第 [25] 至第 [136] 为另一部分；按作者姓氏字母顺序排列。

目录

上　篇
核心词汇分类

第一单元

前缀与前缀词

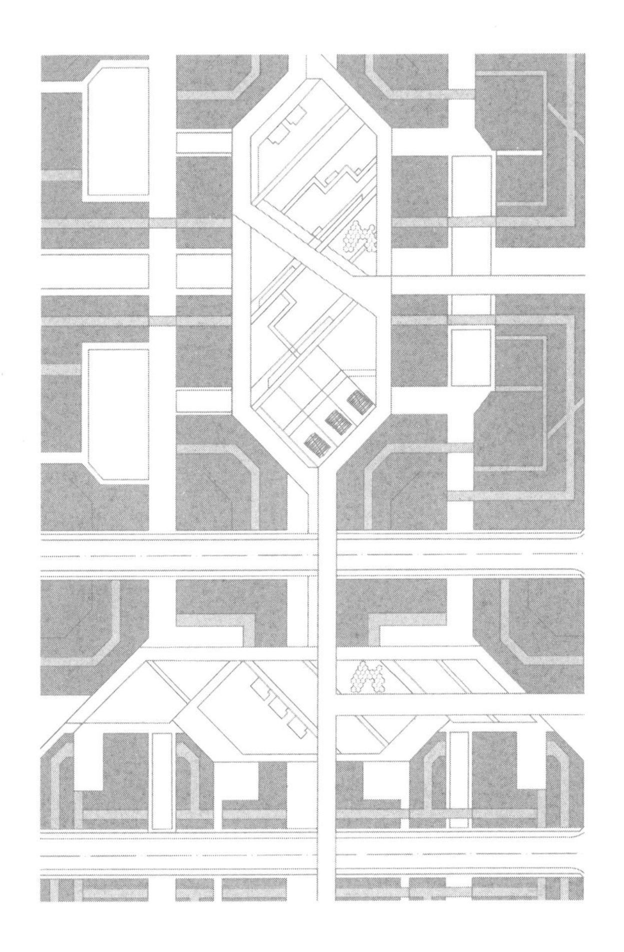

前缀目录

47. out- 过度；外，出	015	63. se- 分离，分开	019
48. over- 过度；在上，在外	016	64. self- 自身的	019
49. paleo- 古，旧	016	65. sub- 下，下面	019
50. pan- 全部	016	66. suc-= sub- 下，下面	019
51. para- 类似，靠近	017	67. sup-= sub- 下，下面	020
52. pen- 几乎，近于	017	68. super- 超过，过分；上面	020
53. per- 贯穿；完全；始终	017	69. sur- 超过	020
54. peri- 周围，环绕	017	70. sym-，syn- 共同，同时	020
55. poly- 多	017	71. trans- 超过；通过；变化	020
56. post- 后	017	72. tri- 三	021
57. pre- 前，预先	018	73. ultra- 超越	021
58. pro- 向前，在前	018	74. un- 不，无，非	021
59. proto- 原始	018	75. under- 低于，在……之下	021
60. pseudo- 假	018	76. uni- 单一	022
61. re- 再次，反复；回，反	018	77. up- 向上，在上	022
62. retro- 向后，回，反	019	78. with- 加强语气	022

前缀与前缀词

1. a-，an- 无，不，非

achromatic adj. 非彩色的，无色的【a-（无）+ chrom（颜色）+ atic →无色的】

apathy n. 漠然，冷淡【a-（无）+ path（感情）+ y →无感情→冷漠，冷淡】

anaerobic adj. 厌氧的；n. 厌氧微生物【an-（不）+ aer（空气）+ obic →不需要空气的】

anarchy n. 无政府，政治上的混乱【an-（无）+ archy（统治）→无统治→无政府】

2. ab- 离开

abdicate v. 退位，辞职，放弃【ab-（离开）+ dic（命令）+ ate →不再命令→退位，辞职】

abnormal adj. 反常的，异常的

abnormally adv. 不正常地

normal adj. 正常的

3. ac-，ad-，af-，ag- 加强语气

acclaim v. 欢呼，称赞 v. 欢呼，称赞【ac-（加强语气）+claim（喊）→欢呼，称赞】

advocacy n. 拥护，支持【ad-（加强语气）+ voc（喊）+ acy →摇旗呐喊→拥护，支持】

afflict v. 使痛苦，折磨【af-（加强语气）+ flict（打击）→使痛苦】

aggrandize v. 增大，扩张【ag-（加强语气）+ grand（大）+ ize →增大】

4. acro- 高

acrobat n. 特技演员，杂技演员【acro-（高）+ bat（打斗）→高空中打斗、高空中表演的人】

acrophobia n. 恐高症【acro-（高）+ phob（恐惧）+ ia →恐高症】

5. ambi-，amphi- 二

ambition n. 抱负，雄心，野心【ambi-（二）+ tion → 二心→野心，雄心】

ambitious adj. 有抱负的，雄心勃勃的

ambivalent adj.（对人或物）有矛盾看法的

amphibian n. 两栖动物；水陆两用飞行器

6. ante- 前，先

antecedence n. 居先，先行【ante-（前，先）+ cede（走）+ nce →走在前面→居先，先行】

antecedent n. 前事，前辈；adj. 先行的

antedate v.（在信、文件上）写上较早日期；早于

7. anti- 反对，相反；前，先

表示"反对，相反"

antipathy n. 反感，厌恶【anti-（反）+ path（感情）+ y →反感，厌恶】

antislavery n. & adj. 反对奴隶制度（的）

anti-social adj. 反社会的

antiwar adj. 反战的

表示"前，先"

antiquarianism n. 古物研究，好古癖

antiquated adj. 陈旧的，过时的

antique adj. 古时的，古老的；n. 古物，古董

antiquity n. 古老；古人；古迹

8. arch- 古代，最先

archaeology n. 考古学

archaeological adj. 考古学的

archeologist n. 考古学家

archaic adj. 古老的

archive n. 档案室

9. auto- 自己；自动

automatic a. 自动的，无意识的；n. 小型自动武器

automation n. 自动化

automobile n. 汽车

autonomous adj. 自治的，自主的

autonomy n. 自治，独立

10. be- 加强语气

befuddlement n. 迷惑不解【fuddle v. 灌醉，使迷糊】

belated adj. 来得太迟的

belittle v. 轻视，贬抑

berate v. 猛烈责骂【rate v. 责骂】

besiege v. 围攻，困扰【siege n. 围攻，围困】

11. bene-，ben- 好

benediction n. 祝福，祈祷【bene-（好）+ dict（说话）+ ion →说好话→祝福，祈祷】

benefactor n. 行善者，捐助者【bene-（好）+ fact（做）+ or →做好事的人→行善者】

beneficent adj. 慈善的，仁爱的；有益的【bene-（好）+ fic（做）+ ent →做好事的】

beneficial adj. 有利的，有益的

beneficiary n. 受益人

benevolent adj. 善心的，仁心的【bene-（好）+ vol（意愿）+ ent →善心的，仁心的】

benign adj. 慈祥的

benison n. 祝福，赐福

12. bi- 二

bidirectional adj. 双向的

biennial adj. 两年一次的【bi-（两）+ enn（年）+ ial →两年一次的】

bifurcate v. 分为两支，分叉

bigamy n. 重婚，重婚罪

biped n. 两足动物【bi-（两）+ ped（脚）→两足动物】

13. by- 旁，侧，副

byline n. 副业；平行干线的铁路支线

byproduct n. 副产品；副作用

bystander n. 旁观者

14. caco- 坏

cacophony n. 难听的声音【caco-(坏)+ phon(声音)+ y →坏声音→难听的声音】

cacophonous adj. 发音不和谐的，不协调的

15. circu-，circum- 环绕，在……周围

circular adj. 圆形的，环形的，循环的

circulate v. 循环，流通；发行

circumlocution n. 迂回累赘的陈述【circum-(环绕)+ locu(说)+ tion →绕圈子说】

circumlocutory adj. 委婉曲折的，迂回的

circumscribe v. 限制，限定【circum-（环绕）+ scribe（写）→画圆圈→画地为牢→限制】

16. co-，col-，com-，con- 共同；加强语气

coexist v. 共存

cooperation n. 合作，协作

collaborate v. 合作,协作;通敌【col-（共同）+ labor（劳动）+ ate →合作,协作】

compose v. 组成;创作;使平静【com-（共同）+ pos（放）+ e →放到一起→组成】

concentric adj. 有同一中心的【con-（共同）+ centr（中心）+ic →有同一中心的】

commute v. 交换;坐公交车上下班【com-（加强语气）+ mut（变化）+ e →交换】

compress v. 压缩，浓缩【com-（加强语气）+ press（压）→压缩】

confront v. 面临，对抗【con-（加强语气）+ front（脸面）→面临，对抗】

consolidate v. 巩固,加强【con-（加强语气）+ solid（结实）+ ate →巩固,加强】

17. contra-，contro-，counter- 反对，相反

contrary adj. 相反的，对抗的

contraband n. 违禁品,走私货【contra-（反）+ band（捆绑,即规矩）→违禁品,走私货】

controversial adj. 引起争论的【contro-（反）+ vers（转）+ ial →反着转→引起争论的】

counteract v. 消除,抵消【counter-(相反)+ act(行动)→相反行动→ 消除,抵消】

countermand v. 撤回（命令），取消（订货）【counter-（反）+ mand（命令）→撤回命令】

18. de- 否定；加强语气

表示“否定”

decompose v.（使）腐烂
deforestation n. 采伐森林
degradation n. 堕落，恶化
deregulate v. 解除管制
devalue v. 贬低，使贬值
表示“加强语气”
delight n. 快乐，高兴，乐事；v. 使高兴，使欣喜
delimit v. 定界，划界
delineate v. 描画

19. di- 两，双
dichotomy n. 两分法；矛盾，分歧；具有两分特征的食物
dilemma n. 困境，左右为难
dioxide n. 二氧化物
diverge v. 分歧，分开
divergent adj. 分叉的，叉开的；发散的，扩散的；不同的

20. dis- 否定；分开，离散
表示“否定”
disabled adj. 残疾的，有缺陷的
disadvantaged adj. 贫穷的，社会地位低下的
disillusion v. 梦想破灭，醒悟
displace v. 移动，移走；替代，取代
disproportionate adj. 不成比例的，不相称的
表示“分开，离散”
discrete adj. 个别的；不连续的
discernible adj. 可识别的，可辨别的
disperse v. 分散，散开，散播
distinguish 区分，辨别

21. du- 二，双
dual adj. 双重的
duel n. 决斗（两人之间的战斗）
duplicate adj. 复制的，两重的；v. 复制；n. 复制品，副本
duplicitous adj. 搞两面派的，奸诈的

duplicity n. 欺骗，口是心非

22. dys– 坏

dysfunction n. 功能障碍，功能失调【function n. 功能，作用】

dysfunctional adj. 功能失调的

dyslexia n. 诵读困难【dys–（坏）+ lex（词语）+ ia →诵读困难】

dyspeptic adj. 消化不良的；不高兴的【peptic adj. 产生胃酶的，助消化的】

23. em– 加强语气；在……中

表示“加强语气”

embitter v. 使痛苦，使难受

embolden v. 壮胆，鼓励

empower v. 授权（给某人）采取行动

表示“在……中”

embay v. 使（船）入海湾

embed v. 牢牢插入，嵌于

24. en– 加强语气；在……中

表示“加强语气”

endanger v. 危及，使遭受危险

enlightenment n. 启迪，启发，教化

enliven v. 使……更活跃

enslave v. 奴役

entitlement n. 授权

表示“在……中”

engulf v. 吞噬【gulf n. 海湾；深渊】

enmesh v. 绊住，陷入网中【mesh n. 网状物，网孔】

25. eu– 好

eugenic adj. 优生（学）的【eu–（好）+ gen（生产）+ ic →优生的】

eulogistic adj. 颂扬的，歌功颂德的【eu–（好）+ log（说）+ istic →说好话的→颂扬的】

eulogy n. 颂词，颂文

euphemism n. 婉言，委婉的说法

euphonious adj. 悦耳的【eu–（好）+ phon（声音）+ ious →好听的声音→悦耳的】

26. ex- 出；外；超过

表示"出"

exhume v. 掘出，发掘【ex-（出）+ hum（地）+ e →从地下挖出→掘出，发掘】

exodus n. 大批离去，成群外出

表示"外"

external adj. 外面的，外部的

externality n. 外在性，外形，外部事物

exotic adj. 珍奇的；来自异国的

表示"超过"

exaggerate v. 夸张；扩大，增加

excess n. 过分，过度

27. extra- 超过

extra adj. 额外的，特别的；adv. 特别地；n. 额外的事物，额外费用

extraordinarily adv. 特别地，不平常地

extravagance n. 奢侈，挥霍

extremist n. 极端主义者

extremity n. 极度；绝境，险境；临终

28. fore- 前，先

forecast v. 预报，预测

forecourt n. 前院

foreknowledge n. 预知

forerunner n. 预兆，前兆；先驱

foreshadow v. 预示

forethought n. 事先的考虑，远见卓识

29. hetero- 异

heterodox adj. 异端的，非正统的

heterogeneity n. 异质性

heterogeneous adj. 异类的，不同的【hetero-（异）+ gene（产生）+ ous →异类的】

heteromorphic adj. 异形的【hetero-（异）+ morph（形状）+ ic →异形的】

30. homo- 同

homogeneity n. 同种，同质

homogeneous adj. 同类的，相似的【homo-（同）+ gene（产生）+ ous →同类的】

homogenize v. 使均匀，使一致

homologous adj. 相应的，一致的

31. hyper- 过分

hype n. 夸大的广告宣传

hyperactivity n. 活动过度，极度亢奋

hyperbole n. 夸张法

hypersensitive adj. 非常敏感的

hypertension n. 高血压【tension n. 紧张，不安】

32. hypo- 下，低

hypocrite n. 伪善者，伪君子【hypo-（下，即背后）+ crit（判断）+ e →在背后评论别人的人→伪善者，伪君子】

hypocrisy n. 伪善，虚伪

hypocritical adj. 虚伪的

33. im- 不，无，非；在……内；加强语气

表示“不，无，非”

immeasurably adv. 不能测量地，无限制地

imperfect adj. 有缺点的，不完美的

impermeable adj. 不可渗透的

impersonal adj. 没有人情味的

impervious adj. 不可渗透的

表示“在……内”

imbue v. 灌输（某人）强烈的情感或意见

imprint v. 盖印，刻印

表示“加强语气”

impoverish v. 使成赤贫【poverty n. 贫穷，贫乏】

34. in- 不，无，非；在……内；加强语气

表示“不，无，非”

incalculable adj. 数不清的，不可估量的

independent adj. 自主的，独立的

inevitable adj. 不可避免的，必然的

inhospitable adj. 不好客的，不友好的，不适于居住的

instability n. 不稳定，不稳固

intangible adj. 触不到的，难以理解的

intolerable adj. 不能忍受的，无法容忍的

invalidate v. 使无效，使作废

invertebrate n. 无脊椎动物；adj. 无脊椎的

表示“在……内”

inclusive adj. 包含一切的，范围广的【in-（在……内）+ clus（关闭）+ ive →包含一切的】

infuse v. 灌输；鼓励【in-（在……内）+ fus（流）+ e →灌输】

表示“加强语气”

inflame v. 使燃烧；激怒（某人）【in-（加强语气）+ flame（火焰）→使燃烧】

35. inter- 在……之间，互相

表示“在……之间”

interdisciplinary adj. 跨学科的

intermediate adj. 中间的，中级的

interpersonal adj. 人与人之间的，人际的

表示“互相”

interaction n. 相互作用，相互影响

interconnected adj. 相互连接的

interrelate v. 相互关联，相互影响

interrelationship n. 相互关系

36. ir- 不，无，非

irrefutable adj. 无可辩驳的，毋庸置疑的

irregular adj. 不规则的，不整齐的

irrelevant adj. 不相关的，不切题的

irresolute adj. 未决定的，犹豫不决的

irreversibly adv. 不可更改地，不可挽回地，不可反转地

irrevocable adj. 无法取消的，不能改变的

37. macro- 大

macro adj. 巨大的

macroeconomics n. 宏观经济学

macrochange n. 大变动

macroplan n. 庞大的计划

38. mal- 坏

maladroit adj. 笨拙的【adroit adj. 熟练的，灵巧的】

malediction n. 诅咒【mal（坏）+ e + dict（说话）+ ion →说坏话→诅咒】

malefactor n. 罪犯，作恶者【mal-（坏）+ e + fact（做）+ or →做坏事的人】

malfunction v. 发生故障；n. 故障【function n. 功能，作用】

malnutrition n. 营养不良

malodor n. 恶臭【odor n. 气味，名声】

malpractice n. 玩忽职守，渎职

39. meta- 变化

metabolism n. 新陈代谢

metamorphose v. 变形【meta-（变化）+ morph（形态）+ ose →变形】

metamorphosis n. 变形

40. micro- 小

microeconomics n. 微观经济学

microbe n. 微生物

microorganism n. 微生物，细菌

organism n. 生物，有机体

microscopic adj. 极小的；显微镜的

41. mis- 坏

misalliance n. 不适当的结合【alliance n. 结盟，联盟，同盟】

misanthrope n. 愤世嫉俗者【mis-（坏）+ anthrop（人）+ e →坏人→愤世嫉俗者】

misbehavior n. 不礼貌，品行不端

misconception n. 误解

mishandle v. 粗暴地对待，错误地处理

42. mono- 单一

monofunctional adj. 单一功能的

monopolize v. 垄断，独占

monopoly n. 专利权，垄断
monoxide n. 一氧化物

43. multi- 多
multibillion adj. 数十亿的
multidimensional adj. 多维的，多面的
multinational adj. 多国的，跨国的
multi-objective adj. 多目标的
multiple adj. 多重的，多个的，复杂的
multitude n. 大量，众多

44. neo- 新
neoclassical adj. 新古典主义的
neolithic adj. 新石器时代的【neo-（新）+ lith（石头）+ ic →新石器时代的】
neologism n. 新字,（旧词的）新义【neo-（新）+ log（话）+ ism →新语言→新字】
neophyte n. 初学者，新手

45. non- 不，无，非
non-compatible adj. 不兼容的，不相容的
non-native adj. 非土著的，非天生的，非本国的
nonporous adj. 无孔的，不渗透的【porous adj. 可渗透的，多孔的】
nonprofit adj. 非营利的
non-recyclable adj. 不可循环再用的

46. omni- 全
omnipotent adj. 全能的，万能的【omni-（全部）+ pot（力量）+ ent →全能的】
omnipotence n. 全能，万能，无限威力
omnipresent adj. 无处不在的【present adj. 出席的，在场的；目前的，现在的】
omniscience n. 全知，全知者，上帝
omniscient adj. 无所不知的，博识的【omni-（全部）+ sci（知道）+ ent →全知道的】
omnivorous adj. 杂食的；兴趣杂的【omni-（全部）+ i + vor（吃）+ ous →杂食的】

47. out- 过度；外，出
表示“过度”
outdated adj. 过时的

outpace v. 超过

outstrip v. 超过，跑过

表示“外，出”

outcast n. 被驱逐者，流浪者

outcry n. 大声疾呼，抗议

outlying adj. 偏僻的，边远的

outward adv. 向外，在外

48. over- 过度；在上，在外

表示“过度”

overconfident adj. 过于自信的，自负的

overcrowd v.（使）过度拥挤

overdue adj. 过期未付的；逾期的

overemphasize v. 过分强调

表示“在上，在外”

overflow v. 溢出；充满

overlap v.（部分地）重叠

overlay v. 覆盖；镀金

overlook v. 忽视，俯视

oversee v. 监督，监视

49. paleo- 古，旧

paleozoology n. 古动物学

paleobotany n. 古植物学

paleoclimate n. 古气候学

paleoanthropology n. 古人类学

50. pan- 全部

panacea n. 万灵药

pandemic adj.（疾病）大范围流行的【pan-（全部）+ dem（人民）+ ic →涉及全部民众的】

panorama n. 概观，全景

panoramic adj. 全景的，全貌的，概论的

pantheon n. 万神殿【pan-（全部）+ the（神）+ on →众神之地→万神殿】

51. para- 类似，靠近

parallel adj. 平行的；类似的；n. 平行线；v. 与……相似

parallelism n. 平行，类似

unparalleled adj. 无比的，空前的

paramount adj. 最重要的，最高权力的【para-（靠近）+ mount（山，峰）→靠近山顶的，靠近最高处的→最重要的，最高权力的】

52. pen- 几乎，近于

peninsula n. 半岛

insular adj. 岛屿的；心胸狭窄的

penultimate adj. 倒数第二的

ultimate adj. 最后的

53. per- 贯穿；完全；始终

permeate v. 扩散，渗透

pervious adj. 可渗透的【per-（贯穿）+ vious → 可渗透的】

perambulate v. 巡视；漫步【per-（完全）+ amble（行走）+ ate →到处走→巡视】

persevere v. 坚持不懈【per-（始终）+ severe（adj. 严格的）→坚持不懈】

perpetuate v. 使永存，使永记不忘【per-（始终）+ pet（追求）+ ate →连续不断的】

54. peri- 周围，环绕

peripatetic adj. 巡游的

peripheral adj. 不重要的，外围的

periphery n. 不重要的部分；外围

periphrastic adj. 迂回的，冗赘的

55. poly- 多

polymath n. 知识广博者

polyglot adj. / n. 通晓多种语言的（人）

polytechnic adj. 有关多种工艺的，工艺教育的

56. post- 后

posterior adj.（在时间、次序上）较后的

posterity n. 后代，子孙

posthumous adj. 死后的，身后的【post-（后）+ hum（地）+ ous →死后埋在地

下→死后的】

postwar adj. 战后的

57. pre- 前，预先

precede v. 在……之前，早于【pre-（前）+ cede（走）→在……之前，早于】

precursor n. 先驱，先兆【pre-（前）+ cur（跑）+ sor →先驱，先兆】

predetermine v. 预先裁定，注定【determine v. 确定，决定】

prefabricate v. 预制【fabricate v. 捏造，伪造；建造，制造】

previously adv. 事先，以前

58. pro- 向前，在前

proceed v. 继续前进,举行【pro-（向前）+ ceed（走）→向前走→继续前进,举行】

proclaim v. 宣告,宣布;显示【pro-（前）+ claim（喊）→在人群前面喊→宣布，宣告】

promote v. 提升；促进【pro-（前）+ mot（动）+ e →向前动→提升，促进】

59. proto- 原始

prototype n. 原型；典型

protozoan n. 原生动物

60. pseudo- 假

pseudoscience n. 伪科学

pseudonym n. 假名，笔名【pseudo-（假）+ onym（名称）→假名，笔名】

pseudo-democratic adj. 假民主的

61. re- 再次，反复；回，反

表示"再次，反复"

rebroadcast v. 重播

redevelopment n. 再开发，再发展

reexamination n. 重考；复试；再检查

refresh v. 消除……的疲劳，使精神振作

renewable adj. 可更新的，可再生的

renewal n. 更新，重建

表示"回，反"

recede v. 后退;收回（诺言等）【re-（回,反）+ cede（走）→向回走→后退,收回】

regress v. 倒退，逆行【re-（回，反）+ gress（走）→倒退】

retort v. 反驳【re-（回，反）+ tort（扭）→反驳】

retract v. 缩回，收回【re-（回，反）+ tract（拖）→缩回，收回】

retreat v. 撤退，后退

62. retro- 向后，回，反

retrospect n. 回顾；v. 回顾，回想【retro-（向后）+ spec（看）+ t →向后看→回顾的】

retrospective adj. 回顾的，追溯的

63. se- 分离，分开

secede v. 正式脱离或退出（组织）

segregate v. 分离，隔离

separation n. 分离，隔离

sever v. 切断，脱离

severance n. 切断，断绝

64. self 自身的

self-contained adj. 独立的，自给自足的

self-defeating adj. 自我挫败的，违背自己利益的，自我拆台的

self-sufficient adj. 自足的，极为自负的

65. sub- 下，下面

subdiscipline n. 学科分支

subdivide v. 再分，细分

subordinate adj. 次要的；下级的；n. 下级【sub-（下）+ ordin（顺序）+ ate →顺序在下的】

subside v.（建筑物）下陷；（天气）平息【sub-（下面）+ sid（坐）+ e →坐下去→下陷】

subterranean adj. 地下的，隐藏的

66. suc-= sub- 下，下面

successively adv. 接连地，继续地【suc-（下，后）+ cess（走）+ ively →在后面跟着走】

succumb v. 屈从；因……死亡【suc-（下面）+ cumb（躺）→躺下→屈从；死亡】

67. sup–= sub– 下，下面

supplant v. 排挤，取代

supplicate v. 恳求，乞求【sup–（下面）+ plic（弯曲）+ ate →双膝跪下→恳求，乞求】

suppliant adj. 恳求的，哀求的；n. 恳求者

supplicant n. 乞求者，恳求者

suppress v. 镇压；抑制

68. super– 超过，过分；上面

super adj. 超级的，极度的，过分的

superb adj. 上乘的，出色的

superior adj. 较高的，较好的，较多的，上等的

supreme adj. 至高的，极度的

supersede v. 淘汰，取代【super–（上面）+ sed（坐）+ e →坐在别人的位子上→取代】

supervise v. 监督，管理【super–（上面）+ vis（看）+ e →从上面看→监督，管理】

69. sur– 超过

surcharge v. 对……收取额外费用；n. 附加费

surfeit n.（食物）过量，过度；v. 使过量

surpass v. 胜过，超过

surplus adj. 过剩的；盈余的

surrealism n. 超现实主义【realism n. 现实主义】

70. sym–，syn– 共同，同时

symbiosis n. 共生，共生关系【sym–（共同）+ bio（生命）+ sis →生活在一起→共生】

symmetry n. 对称，均衡【sym–（共同）+ metr（测量）+ y →两边测量结果相同→对称】

sympathy n. 同情，同情心【sym–（共同）+ path（感情）+ y →同情】

synchronous adj. 同时发生的【syn–（共同）+ chron（时间）+ ous →同时发生的】

synonymous adj. 同义的【syn–（共同）+ onym（名称）+ ous →同名称的→同义的】

synthesize v. 合成

71. trans– 超过；通过；变化

表示“超过”

transgress v. 冒犯，违背【trans-（超过）+ gress（走）→走过了→冒犯，违背】
transcend v. 超越，胜过【trans-（超过）+ scend（爬）→超越】

表示“通过”

transit n. 通过，经过；运输，运输路线，公共交通系统
transmission n. 传送，传播【trans-（通过）+ miss（送）+ ion→传送，传播】
transparent adj. 透明的，清澈的

表示“变化”

transact v. 办理，交易；商议，谈判【trans-（变化）+ act（行动）→办理，交易】
transfer v. 使转移，使调动
transform v. 改变，改观【trans-（变化）+ form（形状）→改变，改观】
transformation n. 转化，转变
transition 过渡，转变，变迁

72. tri- 三

triangle n. 三角形
trilogy n. 三部曲
tripod n. 画架，三脚架

73. ultra- 超越

ultraclean adj. 超净的，特净的
ultramundane adj. 超俗的，世界之外的【mundane adj. 现世的，世俗的】
ultrasonic adj. 超音速的，超声波的【ultra-（超）+ son（声音）+ ic →超音速的】

74. un- 不，无，非

unaffordable adj. 买不起的，负担不起的
unfeasible adj. 不能实行的，难实施的
unfettered adj. 不受限制的，不受约束的
unfold v. 展开，打开
unintentional adj. 无意的，无心的
unresponsive adj. 无答复的，反应迟钝的
unsightly adj. 难看的，不好看的
unsustainable adj. 不能持续的，不能支持的

75. under- 低于，在……之下

underlie v. 位于……之下，构成……的基础

underlying adj. 在下面的；根本的

undermine v. 破坏，损坏【under-（下）+ mine（v. 开矿）→地下采矿→地表受到破坏】

underpinning n. 基础，基础材料，基础结构

underprivileged adj. 贫穷的，下层社会的，被剥夺基本权利的

underrepresented adj. 未被充分代表的

76. uni- 单一

unanimous adj. 全体意见一致的【un（i）-（一）+ anim（精神）+ ous →全体意见一致的】

uniformity n. 均匀性，单调，无变化

unify v. 统一，使成一体；使相同

unique adj. 独一无二的，独特的；无与伦比的

universal adj. 全体的；普遍的

universalizable adj. 可通用的，可普遍应用的

77. up- 向上，在上

update v. 更新，使现代化

upgrade n. 升级，上升，增加；v. 使升级，提升

uphold v. 举起，高举；维护，支持

78. with- 加强语气

withhold v. 扣留，保留

withstand v. 顶住；经受住

withdraw v. 撤退，收回；隐居

第二单元

词根与词根词

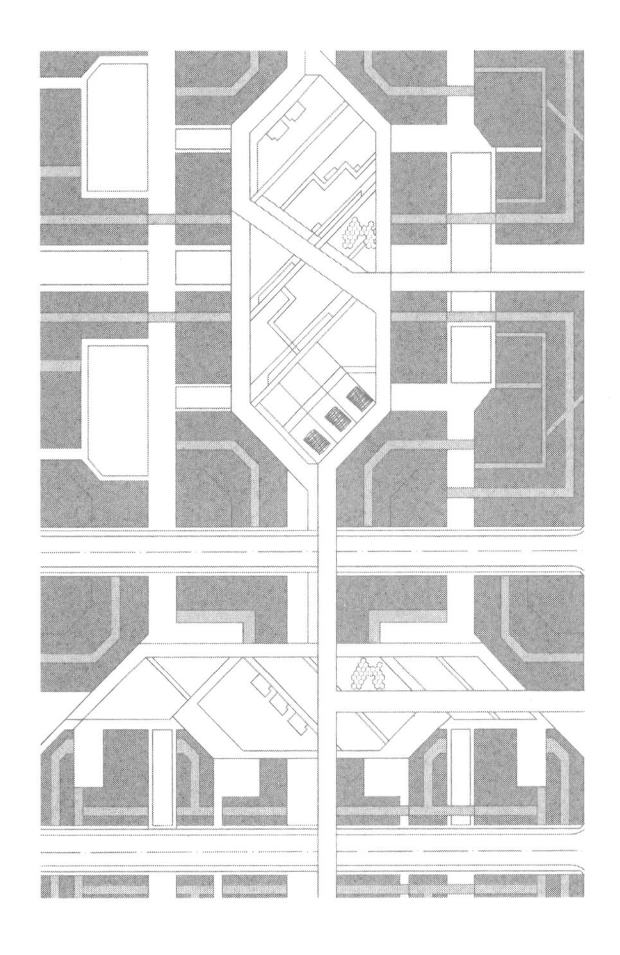

词根目录

45. cre，crue 增加	039	84. graph（y）写，画	050
46. cred 相信	039	85. grav 重	050
47. crim 罪行	039	86. greg 群，团体	050
48. crit 判断	040	87. gress 走	051
49. culp 罪行	040	88. hale 气	051
50. cumb 躺	040	89. herb 草	051
51. cumul 堆积	040	90. here 黏连	051
52. cur，cour 跑	040	91. here 异	052
53. dain，dign 高贵	041	92. hor 恨，怕	052
54. dem 人民	041	93. hum 人	052
55. dict，dic 说话，命令	041	94. hum 地	052
56. duct，duc 引导	042	95. hydr 水	052
57. eco 生态，经济	042	96. icon 形象，偶像	053
58. equ 相等，公平	042	97. insul 岛	053
59. err 错误	043	98. it 走	053
60. ethno 人种，种族	043	99. ject 抛，扔	053
61. fac，fact 做	043	100. jud 判断，审判	053
62. face 脸，面	044	101. junct 结合，连接	054
63. fall 错误	044	102. just 正确	054
64. ferv，ferm 热	044	103. labor 劳动	054
65. fic，fict 做	044	104. laps 滑，滑落	054
66. fid 相信	045	105. leg 法律	055
67. figure 形状，形体	045	106. lev 举，升，使……轻	055
68. fin 范围	046	107. lex，lix 词语	055
69. firm 坚固，坚定	046	108. liber 自由	055
70. flam 火焰，燃烧	046	109. life 生命	056
71. flat 吹气，充气	046	110. liter 字母，文字	056
72. flect，flex 弯曲	047	111. lith 石头	056
73. flict 打击	047	112. locu 说，言	056
74. flor，flour 花	047	113. log，logu 说，言	056
75. flu 流	047	114. loqu 说	057
76. form 形状，形成	047	115. luc 清晰，光	057
77. frac，fract，frag 碎裂	048	116. lumin 光	057
78. front 前面，脸面	048	117. lustr 光	058
79. funct 作用，功能	048	118. magn 大	058
80. fus 流	048	119. man，manu 手	058
81. gene，gen 产生，生产	049	120. mand 命令	058
82. gnos，gnor 知道	049	121. mania 狂热	058
83. grand 大，雄伟	050	122. medi 中间	059

123. memor 记住	059
124. mend 修补，改善	059
125. mens 测量	059
126. meter，metr 仪表，测量	060
127. migr 移动	060
128. mini 小	060
129. miss，mis，mit 投，送，发	060
130. mob 动	061
131. morph 形状，形态	061
132. mot 动	061
133. mount 山，上升	061
134. mun 公共	062
135. mut 改变	062
136. noc，nox，nic 毒，伤害	062
137. nomin，nom 名称，名字	062
138. norm 规则，标准	063
139. not 标示，记号	063
140. nounc，nunc 说出，报告	063
141. nov 新	063
142. numer 数字	064
143. odor 气味，闻	064
144. oner 负担	064
145. onym 名称，名字	064
146. orbit 轨道，常规	064
147. ordin 顺序，正常	064
148. ori 起源，产生	065
149. ortho 直的，正确的	065
150. orn 装饰	065
151. par 平等	065
152. part 部分，分开	066
153. passi，path，pati 感情；疾病	066
154. patri 父亲，国家	067
155. pecc 斑点	067
156. ped 脚	067
157. pel 推，逐，驱	067
158. pend，pens 悬，挂	068
159. pend，pens 花费	068
160. pet 追求	068
161. phil 爱	068
162. phob 憎恨，恐惧	069
163. phon 声音	069
164. photo 光	069
165. pict 描画	069
166. plac 平静，满足	069
167. plen 满	069
168. plet 满	070
169. pli 弯曲	070
170. pos 放	071
171. pot 力量	072
172. press 压	072
173. prim，pri 第一，最早	072
174. priv 私人的	072
175. puls，vuls 推，冲	073
176. qui 平静，安静	073
177. rect 直	073
178. rupt 断，破裂	073
179. scend，scent 攀，爬	074
180. sci 知道	074
181. scribe，script 写，记录	074
182. sect 切	075
183. sed 坐	075
184. semble 和……相像	075
185. semin 种子	075
186. sent，sens 感觉	075
187. sequ 跟随	076
188. serv 保存	076
189. sid 坐	076
190. sign 签名，标记	077
191. simil，simul 像，相同	077
192. sist 站，立	077
193. soci 同伴，社会	078
194. sol 孤单，安慰	078
195. solid 结实	078
196. solv，solu 溶解，松开	078
197. son 声音	078
198. spec 看	079
199. spic 看	079
200. spir 呼吸	080

201. struct 建造，结构	080	219. vad，vas 走	084
202. sum 总，概要	080	220. vag 流浪，漂泊	085
203. surge 上升，起立	080	221. vari 变化	085
204. tach，tact 接触	081	222. ven，vent 来	085
205. techn 技术，技巧	081	223. ver 真实	085
206. tem 节制	081	224. verb 词语	086
207. tempor 时间，时代	081	225. verg 倾向	086
208. tend，tens，tent 拉	082	226. vert，vers 转	086
209. term 边界，结束	082	227. vi 道路	087
210. the 神	082	228. vigor 活力	088
211. tort 扭，弯曲	082	229. vis，vid 看，想	088
212. tox 毒，伤害	083	230. vit 活力，生命	088
213. tract 拉，拖	083	231. viv 活	088
214. tribut 给予，赠予	083	232. voc，vok 叫，喊	089
215. tric 复杂，迷惑	083	233. vol 意志，意愿	089
216. trud，trus 伸	084	234. volu，volv 转	089
217. turb 扰乱	084	235. vor 吃	090
218. vacu 空	084		

词根与词根词

1. acid，acri，acu 尖，酸；敏锐

acid adj. / n. 酸（的）

acidic adj. 酸的，酸性的

acrid adj. 辛辣的，刻薄的

acrimonious adj. 尖酸刻薄的，激烈的

acrimony n. 尖刻，刻薄

acuity n. 敏锐

acumen n. 敏锐，精明

acute adj. 敏锐的，灵敏的；（病）急性的

2. act 行动

activist n. 激进主义分子，积极分子

inactivity n. 不活动，静止

counteract v. 消除，抵消【counter-（相反）+ act（行动）→相反行动→ 消除，抵消】

enact v. 制定（法律）；扮演（角色）

reenact v. 再制定

interaction n. 相互作用，相互影响【inter-（互相）+ act（行动）+ ion → 相互作用】

interactive adj. 相互作用的，相互影响的

proactive adj. 积极主动的【pro-（前）+act（行动）+ ive →向前动→积极主动的】

react v. 反应，作出反应

reaction n. 反应

transact v. 办理，交易；商议，谈判【trans-（变化）+ act（行动）→办理，交易】

3. aer 空气

aerate v. 充气，让空气进入

aerial adj. 空中的，空气中的

anaerobic adj. 厌氧的；n. 厌氧微生物【an-（不）+ aer（空气）+ obic →不需要空气的】

aeronautics n. 航空学

4. ag 动，做

agile adj. 敏捷的，灵活的

agility n. 敏捷

agitate v. 搅动，煽动；使不安，使焦虑

agitated adj. 激动的，不安的

coagulate v. 使凝结【co-（一起，共同）+ ag（做）+ ulate →做到一起→凝结】

coagulant n. 凝结剂，凝血剂

5. agon 挣扎，打斗

agonize v. 使痛苦，折磨【agon（挣扎）+ e →痛苦挣扎→痛苦】

agony n. 极大的痛苦

antagonism n. 对抗，敌对【ant（i）-（反）+agon（打斗）+ ism →对抗，敌对】

antagonistic adj. 敌对的，对抗性的

antagonize v. 使对抗，与……对抗

6. alien 疏远

alien adj. 外国的，异己的；n. 外国人，外星人

alienate v. 疏远，离间某人

alienation n. 疏远，离间

inalienable adj. 不可剥夺的

7. alter，altr 改变；其他

alter v. 改变，更改

alternate adj. 轮流的，交替的；v. 轮流，交替；n. 候选人，替代性选择

alternative adj. 择一的，供选择的；n. 选择，取舍；替换物，选择对象

unalterable adj. 不能改变的

altruism n. 利他主义，无私

altruistic adj. 利他的，无私心的

8. am 爱，情爱

amateur n. 业余爱好者

amiable adj. 和蔼的，亲切的

amicable adj. 友好的

amity n.（人或国）友好关系

enamored adj. 珍爱的，喜爱的【en-（加强语气）+ am（爱）+ ored →珍爱的，喜爱的】

9. amble，ambul 行走，走来走去

amble n. / v. 漫步，缓行

ambulatory adj.（适宜于）步行的

perambulate v. 巡视；漫步【per-（完全）+ amble（行走）+ ate →到处走→巡视】

preamble n. 前言，序言；先兆

ramble n. 漫步；v. 漫步

scramble v. 攀登；争夺

shamble v. 蹒跚而行，踉跄而行

10. anim 生命，精神

animate adj. 活的，有生命的；v. 赋予生命

animated adj. 活生生的，生动的

animation n. 生气，活泼；动画片

inanimate adj. 无生命的

unanimous adj. 全体意见一致的【un（i）-（一）+ anim（精神）+ ous →全体意见一致的】

11. ann，enn 年

annals n. 编年史

annual adj. 每年的，一年生的

biennial adj. 两年一次的【bi-（两）+ enn（年）+ ial →两年一次的】

millennium n. 一千年；（未来的）太平盛世【mill-（千）+ enn（年）+ ium →一千年】

perennial adj. 终年的，永久的【per-（始终）+ enn（年）+ ial →终年的，永久的】

superannuated adj. 老迈的【super-（超）+ ann（年）+ uated →老迈的】

12. anthrop 人

anthropology n. 人类学

anthropologist n. 人类学家

anthropological adj. 人类学的

anthropoid adj. 像人类的；n. 类人猿

anthropogenic adj. 人类起源论的，人为的【anthrop（人）+ o + gen（产生）+ ic →人为的】

anthropometric adj. 人体测量的【anthrop（人）+ o + metr（测量）+ ic →人体测量的】

misanthrope n. 愤世嫉俗者【mis-（坏，不良）+ anthrop（人）+ e →愤世嫉俗者】

philanthropic adj. 博爱的【phil（爱）+ anthrop（人）+ ic →爱人的→博爱的】

philanthropist n. 慈善家

13. aqu 水

aquarium n. 水族馆

aquatic adj. 水生的，水中的

aqueduct n. 引水渠，沟渠，高架渠

aquifer n. 地下蓄水层

aqueous adj. 水的

14. arch（y）统治

anarchy n. 无政府，政治上的混乱【an-（不，无）+ archy（统治）→无统治→无政府】

hierarchy n. 阶层；等级制度

hierarchical adj. 等级的，分层的

monarch n. 君主，帝王【mon（o）-（单个）+ arch（统治）→个人统治→君主，帝王】

monarchy n. 君主制

patriarchal adj. 家长的，族长的；父权制的【patri（父亲）+ arch（统治）+ al →父权制的】

matriarchy n. 母权制，妇女统治【matri（母亲）+ archy（统治）→母权制，妇女统治】

15. art 技巧，艺术

artifact n. 人工制品【art（技巧）+ i + fact（做）→人工制品】

artifice n. 巧办法；诡计【art（技巧）+ i + fic（做）+ e →巧办法】

artificial adj. 人造的，假的

artisan n. 技工

artistry n. 艺术技巧

artless adj. 粗俗的；自然的

16. aster，astr 星星

asterisk n. 星号

asteroid n. 小行星

astrolabe n. 星盘（古代星位观测仪）

astrology n. 占星术，占星学

astronomy n. 天文学

astronomical adj. 天文学的；庞大的

17. avi 鸟

avian n. 鸟；adj. 鸟的，像鸟的，来自鸟类的

aviary n. 鸟舍，鸟类饲养场

aviation n. 航空，航空学，飞行

avifauna n. 鸟类

fauna n. 动物群

18. band，bond 带子；捆绑

abandon v. / n. 放弃；放纵【a-（不）+ band（捆绑）+ on →不再捆绑→放弃，放纵】

band n. 带子；收音机波段

bandage n. 绷带；v. 用绷带包扎

bondage n. 奴役，束缚

contraband n. 违禁品,走私货【contra-（反）+ band（捆绑,即规矩）→违禁品，走私货】

disband v. 解散（团体）【dis-（不）+ band（捆绑）→不再捆绑→解散】

19. bat 打斗

acrobat n. 特技演员，杂技演员【acro-（高）+ bat（打斗）→高空中打斗、高空中表演的人】

aerobatic adj. 飞行技艺的【aer（空气，即高空）+ o + bat（打斗）+ ic →飞行技艺的】

combat n. / v. 搏斗，格斗

combative adj. 好斗的

debatable adj. 未决定的，有争执的

debate n. 正式的辩论，讨论

baton n.（指挥家用的）指挥棒；警棍

battalion n. 军营，军队

20. bell 战斗，战争

bellicose adj. 好战的，好斗的

belligerence n. 交战；好战性，斗争性

belligerent adj. 发动战争的；好斗的，好挑衅的

rebellious adj. 反抗的；难控制的

21. biblio 书

bibliomania n. 藏书癖【biblio（书）+ mania（痴迷）→藏书癖】

bibliography n. 文献学；参考书目【biblio（书）+ graphy（写）→写书用的书→参考书目】

bibliophile n. 爱书者，藏书家【biblio（书）+ phil（爱）+ e →爱书者，藏书家】

22. bio 生命，生物

antibiotic n. 抗生素；adj. 抗菌的【anti-（反）+ bio（生物，即细菌）+ tic →抗菌的】

autobiography n. 自传【auto-（自己）+ bio（生命）+ graphy（写）→写自己的生命→自传】

autobiographical adj. 自传的，自传体的

biodiversity n. 生物多样性

biogas n. 生物气（尤指沼气）

biogeochemical adj. 生物地球化学的

biomass n. 生物量

biome n. 生物群落

biota n. 生物区系，生物群

symbiosis n. 共生，共生关系【sym-（共同）+ bio（生命）+ sis →生活在一起→共生】

23. cad，cid，cas 落下

cascade n. 小瀑布【cas（落）+ cad（落）+ e →小瀑布】

coincide v. 巧合，一致【co-（共同）+ in + cid（落）+ e →共同落下→巧合，一致】

coincidental adj. 巧合的，同时发生的

decadence n. 衰落，颓废【de-（加强语气）+ cad（落）+ ence →衰落】

deciduous adj. 非永久的，短暂的；脱落的，每年落叶的【de-（加强语气）+ cid（落）+ uous →脱落的，落叶的】

occidental n. /adj. 西方（的）【oc + cid（落）+ ental →太阳落下的地方→西方（的）】

24. cant，chant 唱，说

chant n. 圣歌；v. 歌唱或背诵

chantey n. 船歌

enchant v. 使陶醉；施魔法于【en-（加强语气）+ chant（唱）→使陶醉】

disenchant v. 对……不再抱幻想，使清醒

recant v. 改变，放弃（以前的声明或信仰）【re-（回，反）+ cant（说）→原来说的不算了→改变，放弃】

recantation n. 改变宗教信仰

25. cap 头

capita n. 头，首

capital n. 首都；资本，资源；大写字母

capitalize v. 用大写字母写或印刷；使……资本化；把……定为首都

capitation n. 人头税

capitulate v.（有条件地）投降（低头）

recapitulate v. 扼要重述 [重新把重要的东西（即“头”）叙述]

26. cap 拿，抓住

capacious adj. 容量大的，宽敞的

capsule n. 荚；胶囊

caption n. 标题

captious adj. 吹毛求疵的

captivate v. 迷惑，吸引

capture v. 俘获；夺取或赢得

recapture v. 重新捕获

27. card 心

cardiac adj. 心脏的

cardinal adj. 主要的，基本的

cardiologist n. 心脏病专家

cardiovascular adj. 心血管的

28. carn，carr 肉，肉身

carnage n. 大屠杀，残杀

carnivorous adj. 肉食动物的【carn（肉）+ i + vor（吃）+ ous →肉食动物的】

carrion n. 腐肉

incarcerate v. 下狱，监禁【in-（进入）+ car（肉，即人）+ cerate →人进入→下狱，监禁】

29. cav 洞

cavity n.（牙齿等的）洞，腔

concave adj. 凹的

excavate v. 挖掘，挖出【ex-（出）+ cav（洞）+ ate →从洞里挖出来→挖掘，挖出】

cavern n. 大洞穴

30. cede，ceed，cess 走

accede v. 同意【ac-（加强语气）+ cede（走）→走到一起→同意】

access n. 接近，进入，通路；v. 接近，使用

accessible adj. 易达到的；易受影响的

accessibility n. 可达性

antecedence n. 居先，先行【ante-（前，先）+ cede（走）+ nce →走在前面→居先，

先行】

antecedent n. 前事；前辈；adj. 先行的

exceed v. 超过，超出【ex-（超）+ ceed（走）→ 超过，超出】

excess n. 过分，过度

excessive adj. 过度的，过分的

excessively adv. 过度地

incessant adj. 无间断的，连续的【in-（加强语气）+ cess（走）+ ant →一直走→连续的】

precede v. 在……之前，早于【pre-（前）+ cede（走）→在……之前，早于】

precedent adj. 在先的，在前的；n. 先例；判例

unprecedented adj. 前所未有的【un-（不）+ pre-（前）+ cede（走）+ nted →以前没有人走过→前所未有的】

predecessor n. 前任，前辈；原先的东西

procedure n. 程序，手续

proceed v. 继续前进，举行【pro-（向前）+ ceed（走）→向前走→继续前进，举行】

procession n. 行列；前进

recede v. 后退；收回（诺言等）【re-（回，反）+ cede（走）→向回走→后退，收回】

recess n. 壁凹（墙上装柜子等的凹处）；休假

recession n. 经济萧条时期；撤退，退回

recessive adj. 后退的；隐性的（遗传学）

secede v. 正式脱离或退出（组织）【se-（分开）+ cede（走）→走开→正式脱离】

succession n. 演替

successively adv. 接连地，继续地【suc-（下，后）+ cess（走）+ ively →在后面跟着走】

31. celer 速度

accelerate v. 加速，促进

celerity n. 快速，迅速

decelerate v.（使）减速

32. centr 中心

central adj. 中心的，中央的

centralize v. 使集中，成为……的中心

concentrate v. 聚集，浓缩

concentric adj. 有同一中心的【con-（共同）+ centr（中心）+ic →有同一中心的】

decentralize v. 分散，权力下放

eccentric adj. 古怪的，反常的；（圆）没有共同圆心的；n. 古怪的人

egocentric adj. 利己的【ego（自己）+ centr（中心）+ ic →利己的】

33. chrom 颜色

chromatic adj. 彩色的，五彩的

chromosome n. 染色体

achromatic adj. 非彩色的，无色的【a-（无）+ chrom（颜色）+ atic →无色的】

monochromatic adj. 单色的【mono-（单一）+ chrom（颜色）+ atic →单色的】

monochrome adj. 单色的，单色画的

34. chron 时间

anachronistic adj. 时代错误的【an（a）-（不）+ chron（时间）+ istic（ ）→与时间不合拍的→时代错误的】

chronic adj. 慢性的，长期的

chronological adj. 按年代顺序排列的

synchronous adj. 同时发生的【syn-（共同）+ chron（时间）+ ous →同时发生的】

synchronization n. 同步

35. cide，cise 杀，切，切割

concise adj. 简洁的【con-（加强语气）+ cise（切）→把多余的切掉→简洁的】

incise v. 切，切割

incision n. 切口；切割

incisive adj. 一针见血的

incisor n. 门牙

germicide n. 杀菌剂

herbicide n. 除草剂

insecticide n. 杀虫剂

pesticide n. 杀虫剂

36. cip 落下

precipitate v. 加速，促成；adj. 鲁莽的

precipitation n. 降水，降水量

precipitous adj. 陡峭的

precipice n. 悬崖

precipitant n. 沉淀剂

37. cipher 密码

cipher n. 零；无影响力的人；密码

decipher v. 解开（疑团）；破译（密码）

encipher v. 译成密码

indecipherable adj. 无法破译的

undecipherable adj. 难破译的

38. claim，clam 喊，叫喊

acclaim v. 欢呼，称赞【ac-（加强语气）+claim（喊）→欢呼，称赞】

claim v. 要求或索要；n. 声称拥有的权力

clamor n. 吵闹，喧哗

declaim v. 高谈阔论【de-（加强语气）+claim（喊）→高谈阔论】

declamation n. 雄辩，高调

disclaim v. 放弃权利；拒绝承认【dis-（否定）+claim（喊）→放弃权利】

disclaimer n. 否认，拒绝

exclaim v. 惊叫，呼喊

exclamation n. 惊叹词；惊呼【ex-（出）+ claim（喊）→喊出→惊叫，呼喊】

proclaim v. 宣告，宣布；显示【pro-（前）+ claim（喊）→在人群前面喊→宣布】

proclamation n. 宣布，公布

reclaim v. 纠正；开垦（土地）【re-（回，反）+claim（喊）→原来说的不算了→纠正】

reclamation n. 开垦；改造

39. clin 倾斜，弯曲

decline v. 拒绝；变弱，变小；n. 消减

declination n. 倾斜；衰微

disinclination n. 不愿意，不情愿

proclivity n. 倾向

recline v. 斜倚，躺卧

40. clud，clus，claus 关闭

claustrophobic adj. 患幽闭恐惧症的，导致幽闭恐惧症的【claus（关闭）+ tro +

phob（害怕）+ ic →导致幽闭恐惧症的】

conclusive adj. 最后的，结论的；确凿的，消除怀疑的

exclude v. 排斥，排除【ex-（外）+ clud（关闭）+ e →关在外面→排斥，排除】

exclusive adj.（人）孤僻的；（物）专用的

inclusion n. 包含，内含物

occlude v. 使闭塞

preclude v. 避免，排除

recluse n. 隐士；adj. 隐居的

reclusive adj. 隐遁的，隐居的

seclude v. 和别人隔离【se-（离开）+ clud（关闭）+ e →和别人隔离】

seclusion n. 隔离；隔离地

41. cord，core 心，核心

accord v. / n. 同意，一致

concord n. 公约；和睦

cordial adj. 热诚的；n. 兴奋剂

cordiality n. 诚恳，热诚

core n. 果心；核心；v. 去掉某物的中心部分

discord n. 不合，纷争

42. corpor，corp 身体，团体

corporeal adj. 肉体的，身体的；物质的

corporal adj. 肉体的，身体的

corporate adj. 团体的；共同的

corpulent adj. 肥胖的

incorporate v. 合并，并入

incorporation n. 合并，编入

incorporeal adj. 无实体的，非物质的，灵魂的

43. cosm 宇宙

cosmic adj. 宇宙的

cosmopolitan n. 世界主义者，四海为家的人；adj. 世界主义的，世界性的

cosmopolitanism n. 世界性，世界主义

cosmos n. 宇宙

44. cracy，crat 统治

aristocracy n. 贵族，贵族政府，贵族统治

aristocratic adj. 贵族的，贵族化的

autocracy n. 独裁政府【auto-（自己）+ crat（统治）→独裁者】

autocrat n. 独裁者

bureaucracy n. 官僚政治【bureau（政府的局、处）+ cracy（统治）→官僚政治】

plutocracy n. 财阀统治

technocrat n. 技术管理人员【techn（技术）+ o + crat（统治）→统治技术的人】

theocracy n. 神权政治【the（神）+ o + cracy（统治）→神权政治】

45. cre，crue 增加

accrete v. 逐渐增长；连生

accrue v.（利息等）增大，增多

crescendo n.（音乐）渐强；高潮【cre（增加）+ scend（攀，爬）+ o →渐强，高潮】

increment n. 增值，增加【in-（加强语气）+ cre（增加）+ ment →增值，增加】

incremental adj. 增加的

incrementalism n. 渐进主义

incrementally adv. 逐渐地

46. cred 相信

accredit v. 授权【ac-（加强语气）+ cred（相信）+ it →授权】

credence n. 相信，信任

credible adj. 可信的，可靠的

credit n. 赊购；信任；（电影）片头字幕

credo n. 信条

credulous adj. 轻信的，易信的

discredit v. 怀疑；n. 丧失名誉【dis-（否定）+ cred（相信）+ it →怀疑】

incredulity n. 怀疑，不相信【in-（否定）+ cred（相信）+ ulity →怀疑】

47. crim 罪行

discriminate v. 区分；歧视【dis-（分开）+ crim（罪犯）+ inate →区分；歧视】

discriminatory adj. 歧视的，差别对待的

incriminate v. 连累，牵连【in-（进入）+ crim（罪行）+ inate →连累，牵连】

indiscriminate adj. 不加选择的，随意的，任意的

indiscriminately adv. 随意地，任意地

48. crit 判断

criteria n. 评判标准

critic n. 批评家，评论员

critical adj. 批评的；关键的

critically adv. 批判性地，苛求地

criticism n. 评论，批评，批判

criticize v. 评论，批评；挑剔

critique v. & n. 批判，发表评论

hypocrite n. 伪善者，伪君子【hypo-（下，即背后）+ crit（判断）+ e →在背后评论别人的人→伪善者，伪君子】

hypocrisy n. 伪善，虚伪

hypocritical adj. 虚伪的

49. culp 罪行

culpable adj. 有罪的，该受谴责的

culprit n. 犯罪者

exculpate v. 开脱，申明无罪【ex-（出→出来→解脱）+ culp（罪）+ ate →开脱】

inculpate v. 连累；控告；归咎于【in-（进入）+ culp（罪行）+ ate →连累】

50. cumb 躺

cumber v. 拖累，妨碍（躺在路上）

cumbersome adj. 笨重的

encumber v. 妨害，阻碍

incumbent n. 在职者，现任者；adj. 义不容辞的

recumbent adj. 侧卧的；休息的

succumb v. 屈从；因……死亡【suc-（下面）+ cumb（躺）→躺下→屈从；死亡】

unencumbered adj. 无阻碍的

51. cumul 堆积

accumulate v. 积聚，积累

cumulus n. 积云

52. cur，cour 跑

concur v. 意见相同，一致【con-（共同）+ cur（跑）→跑到一起→意见相同】

concurrent adj. 并发的；协作的；一致的

concurrently adv. 同时地

discursive adj. 散漫的，无层次的【dis-（分开）+ cur（跑）+ sive →散漫的】

excursion n. 短途旅游

excursive adj. 离题的；随意的

incur v. 招惹

precursor n. 先驱，先兆【pre-（前）+ curs（跑）+ or →先驱，先兆】

procure v. 取得，获得

procurement n. 取得，获得；采购

scurry v. 急跑，疾行

53. dain，dign 高贵

disdain n. / v. 轻视，鄙视

dignitary n. 显要人物

dignity n. 尊严，尊贵

indignant adj. 愤慨的，愤愤不平的【in-（不）+ dign（高贵）+ ant →因被视不高贵而愤慨】

indignation n. 愤慨

indignity n. 侮辱，轻蔑；侮辱性的言行

54. dem 人民

democratic adj. 民主的【dem（人民）+ o + crat（统治）+ ic →人民统治的→民主的】

democratization n. 民主化

demography n. 人口统计，人口学【dem（人民）+ o + graphy（写）→写出人的多少】

demographic adj. 人口统计的，人口统计学的

pandemic adj. 大范围流行的【pan-（全部）+ dem（人民）+ ic →涉及全部民众的】

55. dict，dic 说话，命令

abdicate v. 退位，辞职，放弃【ab-（离开）+ dic（命令）+ ate →不再命令→退位，辞职】

benediction n. 祝福，祈祷【bene-（好）+ dict（说话）+ ion →说好话→祝福，祈祷】

contradict v. 反驳，驳斥【contra-（反）+ dict（说话）→反驳，驳斥】

dictate v. 口述；命令

dictator n. 独裁者

dictum n. 格言，声明

malediction n. 诅咒【male-（坏）+ dict（说话）+ ion →说坏话→诅咒】

predict v. 预言，预测，预示【pre-（预先）+ dict（说）→预言】

predictive adj. 预言的

predictable adj. 可预知的

verdict n. 判决，决定【ver（真实）+ dict（说）→根据事实说话→判决】

56. duct，duc 引导

abduct v. 绑架，拐走【ab-（离开）+ duct（引导）→绑架，拐走】

adduce v. 给予（理由）；举出（例证）

conduit n. 渠道，引水道；水管

duct n. 管道，槽

ductile adj. 易拉长的，易变形的；可塑的

induce v. 诱导；引起

inducement n. 引诱，引诱物，诱因，动机

inducible adj. 可诱导的

seductive adj. 诱人的

57. eco 生态，经济

eco-city n. 生态城市

ecological adj. 生态的，生态学的

ecologist n. 生态学者，生态学家

ecology n. 生态学

ecosystem n. 生态系统

unecological adj. 非生态的

economic adj. 经济学的，经济（上）的

economical adj. 节省的，节约的；经济学的，经济上的

economics n. 经济学

economise v. 节约，节省

economist n. 经济学家，经济学研究者

economy n. 节约；经济；adj. 经济的

58. equ 相等，公平

equal adj. 相等的，平等的

unequal adj. 不平等的，不均匀的

equalize v. 使相等

equally adv. 平等地，公正地

equate v. 认为……相等或相仿

equation n. 等式；等同，相等

equator n. 赤道

equilibrium n. 平衡

equitable adj. 公平的，公正的

equity n. 公平，公道

equivalent adj. 相等的，相当的；n. 对等物

inequality n. 不平等，不均等

iniquitous adj. 邪恶的，不公正的

iniquity n. 邪恶，不公正

59. err 错误

aberrant adj. 越轨的；异常的

aberration n. 越轨

err v. 犯错误，出错

erroneously adv. 错误地，不正确地

unerringly adv. 无过失地

60. ethno 人种，种族

ethnic adj. 种族的，部落的

ethnocentric adj. 种族优越感的【ethn（种族）+ o + centr（中心）+ ic →种族优越感的】

ethnographic adj. 人种志的【ethn（种族）+ o + graph（写）+ ic →人种志的】

ethnography n. 人种论，民族志

61. fac，fact 做

artifact n. 人工制品【art（技巧）+ i + fact（做）→人工制品】

benefactor n. 行善者，捐助者【bene-（好）+ fact（做）+ or →做好事的人→行善者】

facile adj. 容易做的，肤浅的

facilitate v. 使容易，促进

facilities n.（使事情便利的）设备，工具

facsimile n. 复制本，摹本【fac（做）+ simil（像，相同）+ e →做出相似的东西→复制本】

malefactor n. 罪犯，作恶者【mal-（坏）+ e + fact（做）+ or →做坏事的人→作恶者】

62. face 脸，面

deface v. 损坏

facade n. 建筑物的正面；（虚伪的）外表

facet n.（宝石等的）小平面；侧面

facetious adj. 轻浮的，好开玩笑的

multifaceted adj. 多方面的

preface n. 序言

63. fall 错误

fallacious adj. 欺骗的；谬误的

fallacy n. 谬误，错误

fallibility n. 易于出错，出错性

fallible adj. 会犯错的，易犯错的

64. ferv，ferm 热

ferment v. / n. 发酵；骚动

fermentation n. 发酵

fervent adj. 炙热的，热情的

fervid adj. 炽热的，热情的

fervor n. 热诚，热心

perfervid adj. 非常热心的

65. fic，fict 做

artifice n. 巧办法；诡计【art（技巧）+ i + fice（做）→巧办法】

artificial adj. 人造的，假的

beneficent adj. 慈善的，仁爱的；有益的【bene-（好）+ fic（做）+ ent →做好事的】

beneficiary n. 受益人

deficiency n. 缺陷，不足

deficit n. 不足；赤字

efficacious adj. 有效的

efficacy n. 功效，有效性

efficiency n. 效率，效能

inefficient adj. 无效率的

magnificent adj. 壮丽的，宏伟的；高尚的【magn（大）+ i + fic（做）+ent → 壮丽的】

maleficent adj. 有害的，犯罪的【mal-（坏）+ e + fic（做）+ ent →做坏事的】

munificent adj. 慷慨的；丰厚的【mun（公共）+ i + fic（做）+ ent →为大家做事要慷慨→慷慨的】

proficient adj. 熟练的，精通的【pro-（前）+ fic（做）+ ient →做在别人前面的→熟练的】

suffice v. 足够，（食物）满足【suf-（= sur-，超）+ fic（做）+ e →足够的】

sufficient adj. 足够的

superficial adj. 表面的，肤浅的【super-（上面）+ fic（做）+ ial →做给别人看的→表面的】

66. fid 相信

confidant n. 心腹朋友，知己，密友

confide v. 吐露，倾诉

confidence n. 信任，自信，信心

confidential adj. 机密的

confidently adv. 确信地，肯定地

diffident adj. 缺乏自信的【dif-（= dis-，否定）+ fid（相信）+ ent → 缺乏自信的】

overconfident adj. 过于自信的，自负的

perfidious adj. 不忠的，背信弃义的【per-（远离）+ fid（相信）+ ious →远离相信→不相信→不忠的】

perfidy n. 不忠，背叛

67. figure 形状，形体

configuration n. 结构，配置；轮廓

disfigure v. 毁容

figurative adj. 比喻的，借喻的

figurehead n. 名义领袖；傀儡

prefigure v. 预示；预想

transfigure v. 美化，改观【trans-（变化）+ figure（形状）→美化，改观】

68. fin 范围

affinity n. 密切关系；吸引力

confine v. 限制，禁闭

definite adj. 清楚的，明确的

definitive adj. 明确的，有权威的；最终的

finite adj. 有限的

infinite adj. 无限的，无穷的

infinitesimal adj. 极微小的；n. 极小量

infinity n. 无限的时间或空间

69. firm 坚固，坚定

affirm v. 确认，肯定

affirmation n. 肯定，断言

confirm v. 证实，证明

unconfirmed adj. 未经证实的

infirm adj. 虚弱的

reaffirm v. 再次确定；重申

reaffirmation n. 再肯定

70. flam 火焰，燃烧

enflame v. 使愤怒或激动

flamboyant adj. 艳丽的，炫耀的

flammable adj. 易燃的

nonflammable adj. 不易燃的

inflame v. 使燃烧；激怒（某人）

inflamed adj. 发炎的

inflammation n. 发炎，炎症

71. flat 吹气，充气

conflate v. 合并

deflated adj. 灰心丧气的

flatulence n. 肠胃气胀

flatulent adj. 自负的，浮夸的

inflate v. 使充气，使膨胀

overinflated adj. 过度充气的

72. flect, flex 弯曲

deflect v. 偏离，转向

flexible adj. 易弯曲的，灵活的

flexibility n. 灵活性，机动性，柔韧性

inflexible adj. 坚定的，不屈不挠的

73. flict 打击

afflict v. 使痛苦，折磨

affliction n. 折磨，痛苦；痛苦的原因；灾害

conflict v. / n. 斗争，战斗；冲突，抵触

inflict v. 遭受

infliction n.（强加于人身的）痛苦，刑罚

74. flor, flour 花

flora n. 植物群

floral adj. 花的，植物的

florescence n. 繁花时期

florid adj. 华丽的；（脸）红润的

flourish v. 昌盛，兴旺

75. flu 流

affluent adj. 富裕的，丰富的

effluent n. 污水，工业废水，流出的水流；adj. 流出的，发出的

fluctuate v. 波动，变化

fluid adj. 流体的，流动的；易变的，不固定的

fluvial adj. 河流的，生在河中的

flux n. 流量，流出

influx n. 注入，涌入

superfluity n. 多余的量

superfluous adj. 多余的，累赘的

76. form 形状，形成

conform v. 一致，遵守

conformist n. 遵奉传统者，遵守习俗者

cruciform adj.“十”字形的

formality n. 遵循的规范；拘泥形式；正式

formalized adj. 形式化的，正式的

formation n. 组织，形成；（军队）编队

formulate v. 用公式表述，使公式化；系统地阐述或提出

reformat v. 重定格式

transform v. 改变，变化【trans-（变化）+ form（形状）→改变，改观】

uniform n. 制服；adj. 相同的，一致的

77. frac，fract，frag 碎裂

fraction n. 碎片；小部分

fracture n. 骨折；折断；裂口

fragile adj. 易碎的，易坏的

fragment n. 碎片；分裂

fragmentary adj. 碎片的；片段的

78. front 前面，脸面

affront v. 侮辱，冒犯

confront v. 面临；对抗

confrontation n. 对抗

effrontery n. 厚颜无耻，鲁莽

upfront adj. 坦率的

79. funct 作用，功能

defunct adj. 死亡的【de-（否定）+ funct（功能）→死亡的】

dysfunctional adj. 功能失调的【dys-（坏）+ funct（功能）+ ional →功能失调的】

function v. 运行；n. 功能，职责

functional adj. 起作用的，能运转的；实用的

malfunction v. 发生故障；n. 故障，障碍【mal-（坏）+ funct（功能）+ ion →发生故障】

perfunctory adj. 草率的，敷衍的【per-（远离）+ funct（功能）+ ory →远离功能→不考虑功能→草率的，敷衍的】

80. fus 流

confusion n. 困惑，糊涂；混乱，骚乱

defuse v.（从爆破装置中）卸除引信；缓和紧张状态或危机局面【de-（否定）+

fus（流）+ e →不让流→卸除引信；缓和紧张状态或危机局面】

diffuse v. 散布，（光等）漫射；adj. 漫射的，散漫的；不简洁的

diffusion n. 扩散，弥漫；冗长；反射；漫射【dif-（= dis-，分开）+ fus（流）+ ion →分开流→扩散，弥漫】

infuse v. 灌输；鼓励

profuse adj. 很多的；浪费的

suffuse v.（色彩等）弥漫，染遍

transfuse v. 输血；充满

81. gene，gen 产生，生产

anthropogenic adj. 人类起源论的，人为的【anthrop（人）+o+ gen（产生）+ ic →人为的】

engender v. 产生，引起【en-（加强语气）+ gen（产生）+ der →产生，引起】

eugenic adj. 优生（学）的【eu-（好）+ gen（生产）+ ic →优生的】

gender n. 性

generate v. 造成，产生

regenerate v. 使再生

genesis n. 创始，起源

genetic adj. 遗传的；起源的

genetics n. 遗传学

genuine adj. 真的，真诚的【gen（产生）+ uine →像出生时一样的→真诚的】

heterogeneous adj. 异类的，不同的【hetero-（异）+ gene（产生）+ ous →异类的】

homogeneous adj. 同类的，相似的【homo-（同）+ gene（产生）+ ous →同类的】

indigenous adj. 土产的，本地的；固有的【indi + gen（生产）+ ous →土特产】

ingenious adj. 聪明的，有发明天才的【in + gen（生产）+ ious →聪明是天生的】

ingenuous adj. 纯朴的，单纯的【in + gen（生产）+ uous →像出生时一样的→纯朴的】

disingenuous adj. 不坦率的

82. gnos，gnor 知道

agnostic adj. 不可知论的；n. 不可知论者【a-（不）+ gnos（知道）+ tic →不可知论的】

diagnose v. 判断，诊断

diagnostic adj. 诊断的；n. 诊断

ignorance n. 无知，愚昧

ignorant adj. 无知的，愚昧的

ignore v. 不顾，不理，忽视

prognosticate v. 预测，预示【pro-（前）+ gnos（知道）+ ticate →预测】

prognosis n. 预测，对疾病的发作及结果的预言

83. grand 大，雄伟

aggrandize v. 增大，扩张；吹捧

grandeur n. 壮丽，伟大

grandiloquence n. 豪言壮语，夸张之言【grand（大）+ i + loqu（说）+ ence →豪言壮语】

grandiose adj. 宏伟的；夸大的

grandstand n. 大看台；v. 哗众取宠

84. graph（y）写，画

autobiography n. 自传【auto-（自己）+ bio（生命）+ graphy（写）→写自己的生命→自传】

autobiographical adj. 自传的，自传体的

autograph n. 亲笔稿，手迹；v. 在……上亲笔签名；adj. 亲笔的

calligraphy n. 书法

cartographer n. 绘制地图者

demography n. 人口统计,人口学【dem（人民）+ o + graphy（写）→写出人的多少】

graphic adj. 图表的，生动的

85. grav 重

aggravate v. 加重，恶化

grave adj. 严峻的；n. 墓穴

gravitate v. 被强烈地吸引

gravitational adj. 万有引力的

gravity n. 重力；严肃，庄重

86. greg 群，团体

aggregate adj. 聚集的，合计的；n. 总数，合计；v. 集合，合计

congregate v. 聚集，集合

disaggregate v. 分解

egregious adj.（缺点等）过分的，惊人的

gregarious adj. 群居的；爱社交的

gregariousness n. 群居，合群

87. gress 走

aggression n. 侵略；敌对的情绪或行为【ag-（加强语气）+ gress（走）+ ion → 走到别人的地盘→侵略】

aggressive adj. 好斗的；进取的

aggressively adv. 侵略地，攻击地

aggressor n. 侵略者，攻击者

digress v. 离题【di-（二）+ gress（走）→离题】

digression n. 离题，题外话

digressive adj. 离题的，枝节的

egress n. 出去，出口【e-（e-= ex-，出）+ gress（走）→出去，出口】

ingress n. 进入【in-（在……内）+ gress（走）→进入】

regress v. 使倒退，复原，逆行【re-（回，反）+ gress（走）→倒退】

regressive adj. 退步的，退化的

transgress v. 冒犯，违背

transgression n. 违法，罪过【trans-（超过）+ gress（走）→走过了→冒犯，违背】

88. hale 气

exhale v. 呼出（气）

hale adj. 健壮的，矍铄的（呼吸得很好的）

inhale v. 吸入，吸气

89. herb 草

herbaceous adj. 草本植物的

herbicide n. 除草剂

herbivorous adj. 食草的【herb（草）+ i + vor（吃）+ ous →食草的】

90. here 黏连

adhere v. 黏着；坚持

adherent n. 拥护者，信徒

cohere v. 黏着，附着

coherence n. 连贯，一致

coherent adj. 连贯的，一致的

inhere v. 固有
inherent adj. 固有的，内在的

91. here 异
heresy n. 异端邪说
heretic n. 异教徒
heretical adj. 异端邪说的

92. hor 恨，怕
abhor v. 憎恨，嫌恶
abhorrent adj. 可恨的，讨厌的
horrendous adj. 可怕的，令人恐惧的
horrific adj. 可怕的

93. hum 人
dehumanize v. 使失掉人性
humane adj. 人道的，慈悲的
humanistic adj. 人性的；人文主义的
humanitarian n. 人道主义者
humor v. 纵容，迁就
inhumane adj. 不近人情的

94. hum 地
exhume v. 掘出，发掘【ex-（出）+ hum（地）+ e →从地下挖出→掘出，发掘】
humble adj. 卑微的；v. 使谦卑
humiliate v. 使屈辱
humility n. 谦逊，谦恭
posthumous adj. 死后的，身后的【post-（后）+ hum（地）+ ous →死后埋在地下→死后的】

95. hydr 水
anhydrous adj. 无水的
carbohydrate n. 碳水化合物
dehydrate v. 除去水分，脱水
hydrant n.（消防用的）水龙头；消火栓

hydrate n. 水化物；v. 水化
hydrology n. 水文学，水文地理学

96. icon 形象，偶像
icon n. 圣象，偶像
iconoclast n. 攻击传统观念或风俗的人
iconoclastic adj. 偶像破坏的，打破旧习的
iconography n. 图解，插图，肖像研究

97. insul 岛
insular adj. 岛屿的；心胸狭窄的
insularity n. 岛国状态，与外界隔绝的生活状况；（思想等的）褊狭
insulate v. 使绝缘；使隔离
insulation n. 隔离，孤立；绝缘
peninsula n. 半岛

98. it 走
circuit n. 环行，环行道；线路；电路【circu-（环绕）+ it（走）→环行，环行道】
circuitous adj. 迂回的，绕圈子的
itinerant adj. 巡回的，流动的
itinerary n. 行程表；旅行路线

99. ject 抛，扔
abject adj. 极可怜的；卑下的【ab-（离开）+ ject（抛，扔）→被人抛弃→极可怜的】
dejected adj. 沮丧的，失望的
injection n. 注射；注射剂
objection n. 厌恶，反对
projection n. 凸出物；投掷，发射；预测，规划，设计
subject n. 受支配的人

100. jud 判断，审判
adjudicate v. 充当裁判；判决
judicial adj. 法庭的，法官的
judicious adj. 有判断力的；明智的

judiciousness n. 明智
prejudice n. 偏见，成见；v. 使产生偏见

101. junct 结合，连接
adjunct n. 附加物，附件
conjunction n. 联合；连词
disjunction n. 分离，分裂
disjunctive adj. 分离的；相反的
junction n. 交叉路口；连接
juncture n. 危急关头；接合处

102. just 正确
adjust v. 调整，整理；使适合，适应
adjustment n. 调整，适应
justifiable adj. 有理由的，无可争议的
justification n. 正当理由，好的（正当的）原因；辩护
justify v. 证明……正当
unjustifiable adj. 不合道理的
unjustified adj. 未被证明为正当的，无法解释的
unjustly adv. 不义地，不法地

103. labor 劳动
belabor v. 过分冗长地做或说；痛打
collaborate v. 合作，协作；通敌【col-（共同）+ labor（劳动）+ ate →合作，协作】
collaborative adj. 合作的，协作的
collaborator n. 协作者，合作者
elaborate adj. 精致的，复杂的
elaboration n. 详细的细节，详尽的阐述
labored adj. 吃力的；（文体等）不自然的

104. laps 滑，滑落
collapse v. 坍塌，塌陷；虚脱，晕倒
elapse v. 消逝，过去；n. 消逝
lapse n. 失误；（时间等）流逝
relapse n. / v. 旧病复发；再恶化

105. leg 法律

illegal adj. 违法的

illegitimate adj. 不合法的；私生的

legislate v. 制定法律

legislature n. 立法机关，立法团体

legitimacy n. 合理，合法

legitimate adj. 合法的；正当的

106. lev 举，升，使……轻

alleviate v. 缓和，减轻

elevate v. 将某人或某物举起

lever n. 杠杆；v. 撬动

leverage n. 杠杆作用；力量上的优势

relieve v. 减轻，解除

relieved adj. 宽慰的，如释重负的

107. lex，lix 词语

dyslexia n. 诵读困难【dys-（坏）+ lex（词语）+ ia →诵读困难】

lexical adj. 词汇的；词典的

lexicographer n. 词典编撰人

lexicon n. 词典

prolix adj. 啰嗦的，冗长的

prolixity n. 啰嗦

108. liber 自由

deliberate adj. 深思熟虑的；故意的；v. 慎重考虑【de-（否定）+ liber（自由）+ ate →不是随随便便的→深思熟虑的】

deliberately adv. 故意地，深思熟虑地

deliberation n. 细想，考虑

liberality n. 慷慨；心胸开阔

illiberal adj. 气量狭窄的

liberal adj. 自由的，慷慨的

liberalism n. 自由主义

liberate v. 释放，解放

libertine n. 性行为放纵者，浪荡的人

liberty n. 自由，自由权；随意，冒失

109. life 生命

proliferate v. 繁殖；激增

proliferation n. 增殖，增生

prolific adj. 多产的，多结果的

110. liter 字母，文字

obliterate v. 涂掉，擦掉【ob-（反）+ liter（文字）+ ate →擦掉文字】

literal adj. 字面上的；忠实原义的；精确的

literate adj. 有读写能力的；有文化修养的

literati n. 文人；学者

illiterate adj. 文盲的

unliterary adj. 不矫揉造作的，不咬文嚼字的

111. lith 石头

monolith n. 单块石头；单一的庞大组织

monolithic adj. 巨石的，巨大的【mono-（单一）+ lith（石头）+ ic →一块大石头→巨石的】

neolithic adj. 新石器时代的【neo-（新）+ lith（石头）+ ic →新石器时代的】

paleolithic adj. 旧石器时代的【paleo-（古老的）+ lith（石头）+ ic →旧石器时代的】

112. locu 说，言

circumlocution n. 迂回累赘的陈述【circum-(环绕)+ locu(说)+ tion →绕圈子说】

circumlocutory adj. 委婉曲折的，迂回的

elocution n. 演说的艺术

interlocutor n. 对话者，谈话者

locution n. 语言风格；惯用语

113. log，logu 说，言

apologize v. 道歉；辩解

eulogistic adj. 颂扬的，歌功颂德的【eu-（好）+ log（说）+ istic →说好话的→颂扬的】

eulogize v. 称赞，赞扬

eulogy n. 颂词，颂文

monologue n. 独白；个人长篇演说【mono（单一）+ log（说）+ ue →一个人说话→独白】

neologism n. 新字，（旧词的）新义【neo-（新）+ log（话）+ ism →新语言→新字】

prologue n. 开场白；序幕【pro-（前）+ logu（话）+ e →开场前说的话→开场白】

trilogy n. 三部曲【tri-（三）+ log（说）+ y →三部曲】

114. loqu 说

colloquial adj. 口语的，口头的

colloquium n. 学术讨论会

colloquy n.（非正式的）交谈，会谈

eloquence n. 雄辩，口才

eloquent adj. 雄辩的，流利的

grandiloquence n. 豪言壮语，夸张之言【grand（大）+ i + loqu（说）+ ence →豪言壮语】

loquacious adj. 多嘴的，饶舌的

magniloquent adj. 夸张的【magn（大）+ i + loqu（话）+ent →夸张的】

obloquy n. 大骂，斥责【ob-（反）+ loqu（说）+ y →骂】

soliloquy n. 自言自语；戏剧独白【sol（孤单，独自）+ i + loqu（说）+ y →自言自语】

115. luc 清晰，光

elucidate v. 阐明，说明

lucid adj. 表达清楚的，明白易懂的

pellucid adj. 清晰的，清澈的

translucent adj.（半）透明的【trans-（通过）+ luc（光）+ ent →光能通过的→透明的】

116. lumin 光

illuminate v. 阐明，解释；照亮

illuminati n. 先觉者，智者

luminary n. 杰出人物，名人

117. lustr 光

illustrate v. 为……做插图或图表；说明，阐明

illustrative adj. 解说性的，用作说明的

illustrious adj. 著名的，显赫的

lackluster adj. 无光泽的；呆滞的

luster n. 光辉，光泽；使有光辉，使有光彩；给……增光；发光

lustrous adj. 有光泽的

118. magn 大

magnanimity n. 慷慨

magnanimous adj. 宽宏大量的，慷慨的

magnate n. 财主，巨头

magnificent adj. 壮丽的，宏伟的；高尚的

magnify v. 放大；赞美

magniloquent adj. 夸张的

magnitude n. 重要；星球的光亮度

119. man，manu 手

emancipate v. 解放，解除【e-= ex-（出→出来→解脱）+ man（手）+ cipate →使手解脱，使手自由→解放，解除】

manacle n. 手铐

maneuver v. / n.（军队）调遣；策略，操纵

outmaneuver v. 以策略制胜

manipulate v. 操纵

manipulative adj. 操纵别人的，老于世故的

manual a. 用手的，手工做的；n. 手册，指南

manuscript n. 手稿；手抄本【manu（手）+ script（写）→手稿，手抄本】

120. mand 命令

countermand v. 撤回（命令），取消（订货）【counter-（反）+ mand（命令）→撤回命令】

demand v. 要求，苛求

undemanding adj. 不严格的；要求不高的

mandate n. 命令，指令

mandatory adj. 命令的，强迫的

remand v. 遣回，召回【re-（反）+ mand（命令）→撤回命令】

121. mania 狂热

bibliomania n. 藏书癖

mania n. 癫狂；狂热

megalomania n. 自大狂

monomania n. 偏执狂，狂热病

pyromania n. 纵火狂

122. medi 中间

mediate v. 调停

medieval adj. 中世纪的，中古的

mediocrity n. 平庸，碌碌无为

medium n. 媒介；（细菌等的）生存环境

medium-scale adj. 中等规模的，中型的

intermediary n. 仲裁者；中间物；adj. 中间的，媒介的

123. memor 记住

commemorate v. 纪念（伟人、大事件等）

immemorial adj. 太古的，极古的【im-（不）+ memor（记住）+ ial →时间太久无法记住→太古的】

memento n. 纪念品

memoir n. 回忆录，自传；记事录

memorandum n. 备忘录

memorial n. 纪念碑，纪念物；adj. 纪念的，悼念的

124. mend 修补，改善

amend v. 修正；（通常向更好的方向）变化

amendment n. 改正，修正；修正案

emend v. 订正，校正

mend v. 修改，改进

mendacious adj. 不真的；撒谎的【mend（修补）+ acious →修补得太多→不真的，撒谎的】

mendacity n. 不诚实

125. mens 测量

commensurate adj. 同样大小的；相称的【com-（共同）+ mens（测量）+ urate →测量结果相同→同样大小的】

incommensurate adj. 不成比例的，不相称的

dimension n. 维度，尺寸

unidimensional adj. 一方面的；一维的

immense adj. 极大的；无限的【im-（不）+ mens（测量）+ e →太大无法测量的→巨大的】

immensity n. 巨大之物；无限

126. meter，metr 仪表，测量

anthropometric adj. 人体测量的【anthrop（人）+ o + metr（测量）+ ic →人体测量的】

symmetry n. 对称，均衡【sym-（共同）+ metr（测量）+ y →两边测量结果相同→对称】

symmetrical adj. 对称的

asymmetric adj. 不对称的

geometry n. 几何学

geometrician n. 几何学家

metric n. 度量标准

metrical adj. 测量的；韵律的

odometer n.（汽车）里程表

127. migr 移动

emigrate v. 移居外国【e-（= ex-，外）+ migr（移动）+ ate →移居外国】

immigrate v.（从外国）移来，移居入境【im-（内）+ migr（移动）+ ate →移居入境】

immigrant n. 移民；adj. 移民的

migrate v. 迁移，移动

migratory adj. 迁移的，流浪的

128. mini 小

diminish v.（使）减少，缩小

miniature n. 小画像；缩影

minimize v. 把……减至最低数量或程度

minion n. 奴才，低下之人
minnow n. 小淡水鱼，鲦鱼

129. miss，mis，mit 投，送，发
dismiss v. 解散，解雇
emit v. 放射，发出（光、热、味等）
emission n. 排放，散发
intermission n. 暂停，间歇
missile n. 发射物；导弹
submission n. 从属，服从
submit v. 屈服；提交，呈递
transmit v. 传送，传播【trans-（通过）+ mit（送）→传送，传播】

130. mob 动
demobilize v. 遣散，使复员
immobile adj. 稳定的，不动的，静止的
mobile adj. 易于移动的
mobility n. 可动性，流动性

131. morph 形状，形态
amorphous adj. 无定形的【a-（无）+ morph（形状）+ ous →无形状的】
metamorphose v. 变形【meta-（变化）+ morph（形态）+ ose →变形】
morphology n. 形态学

132. mot 动
automotive adj. 汽车的，自动的
commotion n. 骚动，动乱
demote v. 降级，降职【de-（否定）+ mot（动）+ e →向后动→降级，降职】
locomotion n. 运动，移动
motile adj. 能动的，有自动力的
motility n. 运动性
motivate v. 激发，刺激
motivator n. 促进因素，激发因素
promote v. 提升；促进【pro-（前）+ mot（动）+ e →向前动→提升，促进】
promotive adj. 促进的

remote adj. 遥远的;偏僻的【re-（再,又）+ mot（动）+ e →越动越远→遥远的】

133. mount 山，上升

mountainous adj. 多山的；巨大的

paramount adj. 最重要的，最高权力的【para-（靠近）+ mount（山，峰）→靠近山顶的，靠近最高处的→最重要的，最高权力的】

surmount v. 克服，战胜【sur-（超越）+ mount（山）→克服，战胜】

insurmountable adj. 不能克服的，不能超越的

134. mun 公共

communal adj. 公民的，公共的

commune n. 公社；v. 与某人亲切地交谈

communication n. 通信；交流

communicative adj. 爱说话的，好社交的

municipal adj. 市的，市政的

municipality n. 市，市政当局

munificence n. 慷慨，宽宏大量

munificent adj. 慷慨的；丰厚的【mun（公共）+ i + fic（做）+ cent →为大家做事要慷慨→慷慨的】

135. mut 改变

commute v. 交换；坐公交车上下班

commuter n. 乘公交车辆上下班者，经常乘车往返者

immutability n. 不变，不变性

mutable adj. 可变的；易变的

mutate v. 变异

mutineer n. 反叛者，背叛者

mutinous adj. 叛变的；反抗的

transmute v. 变化，变作【trans-（变化）+ mut（改变）+ e →变化】

136. noc，nox，nic 毒，伤害

innocence n. 无辜，清白

innocuous adj.（行为、言论等）无害的

noxious adj. 有害的，有毒的

obnoxious adj. 令人不愉快的；可憎的

pernicious adj. 有毒的，致命的

137. nomin，nom 名称，名字

denominate v. 命名，取名

denomination n. 命名；（长度、币值的）单位

misnomer n. 名字的误用【mis-（错）+ nom（名称，名字）+ er →名字的误用】

nominal adj. 名义上的，有名无实的

nominate v. 提名；任命，指定

138. norm 规则，标准

enormous adj. 极大的，巨大的

norm n. 规范，准则

normal adj. 正常的，正规的

normative adj. 标准的，规范的

139. not 标示，记号

annotate v. 注解

connotation n. 言外之意；内涵

denote v. 显示，指示

notable adj. 明显的，出众的，重要的

notate v. 以符号表示

notoriety n. 臭名昭著；臭名昭著的人

notorious adj. 臭名昭著的

140. nounc，nunc 说出，报告

announce v. 宣布，发表；通报……的到来

denounce v. 指责【de-（否定）+ nounc（说）+ e →指责】

renounce v.（正式）放弃【re-（回，反）+ nounc（说）+ e →原来说的不算了→放弃】

denunciate v. 公开指责，公然抨击，谴责

enunciate v. 发音；（清楚地）表达，阐明

renunciate v. 放弃

141. nov 新

innovative adj. 革新的

innovation n. 改革，创新

innovative adj. 革新的，创新的

nova n. 新星

novelty n. 新奇，新奇的事物

novice n. 生手，新手

renovate v. 修复，装修，翻新

supernova n. 超新星

142. numer 数字

enumerate v. 列举，枚举【e-（= ex-，出）+ numer（数字）+ ate→把数字说出来→列举】

numerology n. 数字命理学（通过数字算命）

numerous adj. 许多的，很多的

143. odor 气味，闻

malodor n. 恶臭【mal-（坏）+ odor（气味）→恶臭】

odoriferous adj. 有气味的

odorless adj. 无臭的，没有气味的

144. oner 负担

exonerate v. 免除责任；宣布无罪【ex-（出→出来→解脱）+ oner（负担）+ ate→解脱负担→免除责任】

exoneration n. 免除（责任、义务、苦难等）

onerous adj. 繁重的，麻烦的

145. onym 名称，名字

anonymity n. 无名，匿名【an-（不，无）+ onym（名称）+ ity→无名，匿名】

anonymous adj. 无名的，匿名的

eponymous adj. 齐名的

pseudonym n. 假名，笔名【pseudo-（假）+ onym（名称）→假名，笔名】

synonymous adj. 同义的【syn-（共同）+ onym（名称）+ ous→同名称的→同义的】

146. orbit 轨道，常规

exorbitant adj. 过度的，过分的【ex-（超）+ orbit（常规）+ ant→超出常规→过度的】

orbital adj. 轨道的

147. ordin 顺序，正常

ordinary adj. 普通的，平常的；拙劣的，质量差的

coordinate v. 使协调，使调和

uncoordinated adj. 不协调的

inordinate adj. 过度的，过分的【in-（不）+ ordin（正常）+ ate →不正常的→过度的】

subordinate adj. 次要的；下级的；n. 下级【sub-（下面）+ ordin（顺序）+ ate →顺序在下的→次要的】

insubordinate adj. 不服从的，违抗的

148. ori 起源，产生

aboriginal n. 原始居民，土著

orient v. 标定方向，使……向东方；adj. 东方的，新生的；n. 东方【ori（起源）+ ent →太阳升起的地方→东方】

orientation n. 方向，定位

origin n. 起源，根源；出身

original adj. 最初的，原始的；有创意的

originality n. 创造性，独特性

originate v. 发源，发起；创始，发明

149. ortho 直的，正确的

orthodox adj. 正统的

orthodoxy n. 正统，正教

orthogonal adj. 直角的，直交的，矩形的

orthodontics n. 畸齿矫正学

150. orn 装饰

adorn v. 装饰

unadorned adj. 未装饰的，朴素的

ornamental adj. 装饰性的

ornate adj. 华美的；充满装饰的

suborn v. 收买，贿赂【sub-（下面）+ orn（装饰）→私下给人好处→收买，贿赂】

151. par 平等

comparison n. 比较，对照；比喻

disparage v. 贬抑，轻蔑【dis-（否定）+ par（相等）+ age→把对方看得不相等→轻蔑】

disparate adj. 迥然不同的【dis-（否定）+ par（相等）+ ate→各个都不相等→迥然不同的】

disparity n. 不同，差异

parity n.（水平、地位、数量等的）同等，相等

152. part 部分，分开

compartment n. 隔间；车厢

counterpart n. 相对应或具有相同功能的人或物

partial adj. 局部的；偏袒的

partiality n. 偏袒，偏心

impartial adj. 公平的，无私的

impartiality n. 公平，公正

partisan n. 党派支持者；党徒

partition n. 隔开；隔墙

153. passi，path，pati 感情；疾病

compassion n. 同情，怜悯

compassionate adj. 有同情心的

dispassionate adj. 平心静气的

impassioned adj. 慷慨激昂的

unimpassioned adj. 没有激情的

impassive adj. 无动于衷的，冷漠的

passionate adj. 充满激情的

antipathy n. 反感，厌恶【anti-（反）+ path（感情）+ y→反感，厌恶】

apathy n. 漠然，冷淡【a-（无）+ path（感情）+ y→无感情→冷漠，冷淡】

empathy n. 同感，感情移入；全神贯注【em-（加强语气）+ path（感情）+ y→同情】

sympathy n. 同情，同情心【sym-（共同）+ path（感情）+ y→同情】

pathogen n. 病原体

pathological adj. 病态的，不理智的；病理学的

pathology n. 病理学

compatible adj. 和谐共处的，相容的【com-（共同）+ pati（感情）+ ble →有共同感情的→相容的】

compatibility n. 和睦相处，适合，通用性

incompatible adj. 不能和谐共存的

154. patri 父亲，国家

compatriot n. 同胞，同国人

expatriate v. 驱逐出国；脱离国籍

repatriate v.（自异国）遣返

patriarchal adj. 家长的，族长的；父权制的【patri（父亲）+ arch（统治）+ al →父权制的】

patriot n. 爱国者，爱国主义者

patriotism n. 爱国主义，爱国心

patron n. 赞助人，资助人，（宗）守护神

patronage n. 赞助，惠顾

patronize v. 以高人一等的态度对待；光顾，惠顾

155. pecc 斑点

impeccable adj. 无瑕疵的

peccadillo n. 小过失

156. ped 脚

biped n. 两足动物【bi-（两个）+ ped（足，脚）→两足动物】

expeditious adj. 迅速的，敏捷的

impede v. 妨碍【im-（在……中）+ ped（脚）+ e →插一脚→妨碍】

impediment n. 妨碍；障碍物

pedal n. 踏板，脚蹬；v. 骑脚踏车

pedestrian adj. 徒步的；缺乏想象的；n. 行人

157. pel 推，逐，驱

compel v. 强迫【com-（加强语气）+ pel（推）→被人推着→强迫】

compelling adj. 引起兴趣的

dispel v. 驱散，消除【dis-（分开）+ pel（推）→驱散】

expel v. 排出；开除【ex-（外）+ pel（推）→开除】

impel v. 推进；驱使【im-（内）+ pel（推）→推进】

propel v. 推进【pro-（前）+ pel（推）→推进】

repel v. 击退；使……反感【re-（回，反）+ pel（推）→击退】

repellent adj. 令人厌恶的

158. pend，pens 悬，挂

appendage n. 附加物【ap + pend（挂）age →挂上去的东西→附加物】

penchant n. 爱好，嗜好【pen（= pend，挂）+ chant →挂念→老想着→爱好，嗜好】

pendent adj. 吊着的，悬挂的

pendulous adj. 下垂的

pendulum n. 摆，钟摆

perpendicular adj. 垂直的

propensity n. 嗜好，习性

suspend v. 暂缓，中止；吊，悬

159. pend，pens 花费

compensate v. 补偿，赔偿

compensatory adj. 补偿性的，报酬的

expend v. 花费；用光

expenditure n. 消耗，支出

inexpensive adj. 廉价的，便宜的

pension n. 养老金，退休金

recompense v. 报酬，赔偿

160. pet 追求

appetite n. 欲望，食欲；爱好

appetizer n. 开胃品

appetizing adj. 美味可口的，促进食欲的

competing adj. 有竞争性的；不相上下的

perpetuate v. 使永存，使永记不忘【per-（始终）+ pet（追求）+uate →连续不断的】

perpetual adj. 连续不断的；永久的

petition n. 请愿；请愿书

petitioner n. 请愿人

161. phil 爱

bibliophile n. 爱书者，藏书家【biblio（书）+ phil（爱）+ e →爱书者，藏书家】

philanthropic adj. 博爱的【phil（爱）+ anthrop（人）+ ic →爱人的→博爱的】

162. phob 憎恨，恐惧

acrophobia n. 恐高症【acro（高）+ phob（恐惧）+ ia →恐高症】

claustrophobic adj. 患幽闭恐惧症的，导致幽闭恐惧症的【claus（关闭）+ tro + phob（害怕）+ ic →导致幽闭恐惧症的】

phobia n. 恐惧症

xenophobe n. 惧外者【xeno（外国人）+ phob（恐惧）+ e →惧外者】

xenophobia n. 仇外，排外

163. phon 声音

cacophony n. 难听的声音【caco-（坏）+ phon（声音）+ y →难听的声音】

cacophonous adj. 发音不和谐的，不协调的

euphonious adj. 悦耳的【eu-（好）+ phon（声音）+ ious →好听的声音→悦耳的】

phonetic adj. 语音的

symphony n. 交响乐，交响曲【sym-（共同）+ phon（声音）+ y →交响乐】

164. photo 光

photosynthetic adj. 光合作用的，促进光合作用的

photovoltaic adj. 光电池的

165. pict 描画

depict v. 描绘，描画

pictorial adj. 绘画的；有图片的，用图片表示的

picturesque adj. 如画的；独特的，别具风格的

166. plac 平静，满足

complacence n. 自满

complacency n. 满足，安心

complacent adj. 自满的，得意的

placate v. 抚慰，平息（愤怒）

placebo n. 安慰剂

placid adj. 安静的，平和的

167. plen 满

plenary adj. 充分的；全体出席的

plenitude n. 完全；大量

plentitude n. 充分

pleonastic adj. 冗言的【pleon（= plen，满，多）+ astic →冗言的】

replenish v. 补充，再装满

168. plet 满

deplete v. 大量减少；耗尽，枯竭

depletion n. 消耗，用尽

incomplete adj. 不完全的，不完整的

plethora n. 过量，过剩

replete adj. 饱满的，塞满的

169. pli 弯曲

compliance n. 顺从，遵从【com（加强语气）+ pli（弯曲）+ ance →顺从（弯曲即顺从）】

compliant adj. 服从的，顺从的

pliable adj. 易弯的，柔软的

pliant adj. 易受影响的；易弯的

complicate v. 使某事复杂化【com（加强语气）+ pli（弯曲）+ cate →复杂（弯曲即复杂）】

duplicate adj. 复制的，两重的；v. 复制；n. 复制品，副本【du–，前缀，二】

duplicitous adj. 搞两面派的，奸诈的

duplicity n. 欺骗，口是心非

explicate v. 详细解说【ex–（出）+ plic（弯曲）+ ate →从弯曲中走出来→使明白→解释】

explicable adj. 可解释的

inexplicable adj. 无法解释的

explicit adj. 清楚明确的；成熟的，形成的【ex–（出）+ plic（弯曲）+ it →从弯曲中走出来→清楚明确的】

explicitly adv. 明白地，明确地

implicate v. 暗示；牵连

implication n. 暗示；牵连

implicit adj. 暗示的，含蓄的【im–（加强语气）+ pli（弯曲）+ cit →暗示的（弯

曲即暗示）】

imply v. 暗示，暗指

replicate v. 复制

replica n. 复制品

supplicate v. 恳求，乞求【sup-（下面）+ pli（弯曲）+ cate →双膝跪下→恳求，乞求】

suppliant adj. 恳求的，哀求的；n. 恳求者

supplicant n. 乞求者，恳求者

170. pos 放

compose v. 组成；创作；使平静【com-（共同）+ pos（放）+ e →放到一起→组成】

composed adj. 镇定的，沉着的

composure n. 镇静，沉着

composition n. 组成，构成

compost n. 堆肥，混合肥料；v. 把……堆制成堆肥

decompose v.（使）腐烂

discompose v. 使失态，慌张

recompose v. 重写；重新安排

depose v. 免职【de-（否定）+ pos（放）+ e →向下放→免职】

deposit n. 储蓄，存款；沉淀物；v. 存放；使淤积

deposition n. 免职；沉淀；作证

dispose v. 配置，布置；处理，处置；使倾向于，使有意于【dis-（分开）+ pos（放）+ e →分开放→布置，处置】

disposal n. 清除，处理

disposed adj. 愿意的，想干的

disposition n. 处理；天性，气质

exposition n. 阐释；博览会

exposure n. 暴露，显露，曝光【ex-（出）+ pos（放）+ ure →放出来→暴露，曝光】

overexposure n. 过分暴露；（照片）曝光过度

impose v. 征（税）；强加于

imposition n. 强加，被迫接受

interpose v. 置于……之间；介入【inter-（在……中间）+ pos（放）+ e →置于……之间，介入】

oppose v. 反对【op-（反）+ pos（放）+ e →反着放→反对】

opposite adj. 相反的，对立的

repose n. / v. 躺着休息，安睡

transpose v. 变换位置，调换【trans-（变换）+ pos（放）+ e →变换位置】

171. pot 力量

despot n. 暴君

omnipotent adj. 全能的，万能的【omni-（全部）+ pot（力量）+ ent →全能的】

potent adj. 强有力的，有影响力的

potential adj. 潜在的，有可能性的

172. press 压

compress v. 压缩，浓缩

depressant adj. 有镇静作用的；n. 镇静剂

depress v. 使沮丧，使不景气；降低，减少

depressed adj. 消沉的；凹陷的

impressed adj. 被打动的；被感动的

impression n. 印象，感想；盖印，压痕

impressionable adj. 易受影响的

impressive adj. 给人印象深刻的，感人的

oppress v. 压迫，压制

oppressive adj. 高压的，压制性的

press v. 挤压

pressing adj. 紧迫的，迫切的；恳切要求的

repress v. 抑制，压抑；镇压

repressive adj. 抑制的；镇压的，残暴的

suppress v. 镇压；抑制

173. prim，pri 第一，最早

primacy n. 首位，第一位；卓越

primal adj. 原始的，最初的；首要的

primate n. 灵长类（动物）

prime adj. 最初的，原始的；最好的，第一流的

primitive adj. 原始的，远古的；基本的

primordial adj. 原始的，最初的

pristine adj. 太古的；纯洁的；新鲜的

174. priv 私人的

deprivation n. 剥夺；缺乏

deprive v. 剥夺，使丧失【de-（否定）+ priv（私人的）+ e →剥夺】

privacy n. 隐居，隐退；隐私；秘密

privation n. 丧失；贫困

privilege n. 特权，特别利益

privy adj. 个人的；秘密参与的

175. puls，vuls 推，冲【参阅“157. pel 推，逐，驱”】

compulsion n. 强迫；难以抗拒的冲动

compulsory adj. 强制性的，命令性的

expulsion n. 驱逐，开除

impulse n. 冲动；刺激

impulsive adj. 冲动的，由冲动引起的；易冲动的

propulsion n. 推进力

repulse v. 驱逐，击退；厌恶；n. 回绝，拒绝；击退

repulsion n. 厌恶，反感；排斥力

revulsion n. 厌恶，憎恶；剧烈反应

176. qui 平静，安静

acquiesce v. 勉强同意，默许

acquiescence n. 默许

acquiescent adj. 默认的

quiescence n. 静止，沉寂

quiescent adj. 不动的，静止的

177. rect 直

erect adj. 竖立的，笔直的，直立的

rectangle n. 长方形，矩形【rect（直）+ angl（角）+ e →都是直角→长方形，矩形】

rectilinear adj. 直线的

rectify v. 改正，调正；提纯【rect（直）+ ify →使……直→改正，调正】

rectitude n. 正直；公正

resurrect v. 使复活；复兴【re-（再，又）+ sur + rect（直）→再次直立→复活，复兴】

178. rupt 断，破裂

abrupt adj. 突然的，意外的；唐突的

bankrupt adj. 破产的

corrupt adj. 堕落的，腐败的；文体有错误的

incorruptible adj.（道德上）不受腐蚀的

disrupt v. 分裂，瓦解；弄乱，扰乱

disruptive adj. 制造混乱的

erupt v. 爆发；喷出（熔岩、水等）

interrupt v. 暂时中止；打断，打扰

rupture n. / v. 破裂，断裂

179. scend，scent 攀，爬

ascendancy n. 统治权，支配力量

ascent n. 上升，攀登；上坡路；提高，提升

condescend v. 轻慢，怀着优越态度对待人

crescendo n.（音乐）渐强；高潮【cre（增加）+ scend（攀，爬）+ o →渐强】

descend v. 下降，降落；屈尊

descendant n. 后代，后裔

transcend v. 超越，胜过【trans-（超过）+ scend（爬）→超越】

180. sci 知道

consciousness n. 意识，观念；清醒状态；知觉

self-conscious adj. 自觉的；害羞的，不自然的

unconscious adj. 不省人事的；未意识到的

omniscient adj. 无所不知的，博识的【omni-(全部)+ sci(知道)+ ent →全知道的】

prescience n. 预知，先见【pre-（前）+ sci（知道）+ ence →预知，先见】

prescient adj. 有先见的，预知的

181. scribe，script 写，记录

circumscribe v. 限制【circum-（环绕）+ scribe（写）→画圆圈→画地为牢→限制】

manuscript n. 手稿；手抄本【manu（手）+ script（写）→手稿，手抄本】

prescribe v. 开处方；规定

prescription n. 处方；指示，规定

prescriptive adj. 规定的，指定的，命令的

proscribe v. 禁止

proscription n. 禁止，剥夺权利

subscribe v. 签名，签字；同意，赞成

transcribe v. 抄写，转录

182. sect 切

dissect v. 仔细分析，剖析

dissection n. 解剖，剖析

intersect v. 相交，交叉

intersection n. 交叉，相交，交叉点

183. sed 坐

sedate adj. 镇静的

sedative adj.（药物）镇静的；n. 镇静剂

sedentary adj. 久坐的

sediment n. 沉淀物，渣

sedimentary adj. 沉淀的，沉淀性的

sedulity n. 勤奋，勤勉

sedulous adj. 聚精会神的，勤勉的【sed（坐）+ ulous →能够坐得住的→勤奋的，勤勉的】

supersede v. 淘汰，取代【super-（上面）+ sed（坐）+ e →坐在别人的位子上→取代】

184. semble 和……相像

assemble v. 集合，聚集，收集；装配，安装

dissemble v. 隐藏，掩饰（感受、意图）【dis-（否定）+ semble（像）→使不相像（本来相像）→隐藏，掩饰】

ensemble n. 全体，整体；大合唱

resemble v. 与……相似，像

semblance n. 外貌；相似

185. semin 种子

disseminate v. 散布，传播【dis-（分开）+ semin（种子）+ ate →散布，传播】

seminal adj. 有创意的

186. sent, sens 感觉

assent v. 同意，赞成【as-（加强语气）+ sent（感觉）→感觉一致→同意】

consent v. 同意，允许【con-（共同）+ sent（感觉）→感觉相同→意见一致】

consensus n. 意见一致

dissent v. 不同意，持异议【dis-（否定）+ sent（感觉）→感觉不同→不同意】

dissenter n. 持异议者，持不同政见者

resent v. 憎恶，愤恨【re-（反）+ sent（感觉）→憎恨】

resentful adj. 愤恨的，怨恨的

sensation n. 知觉；轰动（的事）

sensible adj. 明智的；可感觉到的

sensitive adj. 敏感的

sensitivity n. 敏感，敏感性

hypersensitive adj. 非常敏感的

sentiment n. 多愁善感；思想感情

presentiment n. 预感，预觉

187. sequ 跟随

obsequious adj. 逢迎的，谄媚的

obsequiousness n. 谄媚

sequacious adj. 盲从的

sequela n. 后继者；后遗症

sequence n. 序列，顺序，连续

sequential adj. 连续的，一连串的

subsequent adj. 随后的，后来的【sub-（下面）+ sequ（跟随）+ ent →随后的】

188. serv 保存

conserve v. 保护，保存，保藏

conservationist n. 自然资源保护者，生态环境保护者

conservative adj. 保守的，守旧的

preserve v. 保护，保持，保存

preservative adj. 防腐的；n. 防腐剂

reserve n. 储备；缄默，谨慎；v. 储备，预定

unreserved adj. 无限制的；未被约定的

189. sid 坐

assiduous adj. 勤勉的；专心的【as + sid（坐）+ uous →能够坐得住的→勤勉的，专心的】

dissident v. 唱反调者【dis-（分开，离散）+ sid（坐）+ ent →分开坐→唱反调者】

preside v. 担任主席，负责，指挥【pre-（前）+ sid（坐）+ e →坐在前面→担任主席】

subside v.（建筑物）下陷；（天气）平息【sub-（下面）+ sid（坐）+ e →坐下去→下陷】

subsidiary adj. 辅助的，次要的

190. sign 签名，标记

designate v. 指明，指出；任命，指派

designation n. 指定；名称，称呼

insignia n. 徽章，袖章

signal n. 信号；v. 发信号；adj. 显著的

signatory n. 签署者，签署国

signature n. 签名，签字

signify v. 表示；有重要性

191. simil，simul 像，相同

assimilate v. 同化，吸收

dissimilar adj. 不同的，不相似的

dissimulate v. 隐藏，掩饰【dis-（否定）+ simul（像）→使不相像（本来相像）→隐藏，掩饰】

facsimile n. 复制本，摹本【fac（做）+ simil（相同）+ e →做出相同的东西→复制本】

simulate v. 假装，模仿

simultaneous adj. 同时发生的

verisimilitude n. 逼真【ver（真实）+ i + simil（像）+ itude →像真的→逼真】

verisimilar adj. 好像真实的；可能的

192. sist 站，立

desist v. 停止【de-（加强语气）+ sist（站）→站住→停止】

persist v. 坚持不懈，执意【per-（始终）+ sist（站）→坚持不懈】

persistence n. 坚持不懈，执意，持续

subsist v. 生存下去；继续存在，维持生活【sub-（下面）+ sist（站）→站下去，坚持下去】

subsistence n. 生存，生计；存在

193. soci 同伴，社会

associate adj. 联合的；n. 合伙人；v. 将人或事物联系起来

dissociate v. 分离，游离，分裂

dissociation n. 分离，脱离关系

sociable adj. 好交际的，友好的，合群的

194. sol 孤单，安慰

console v. 安慰，抚慰

inconsolable adj. 无法慰藉的，悲痛欲绝的

desolate adj. 荒凉的，被遗弃的

isolate v. 孤立，将……从其种群中隔离

solace n. 安慰，抚慰

soliloquy n. 自言自语；戏剧独白【sol（孤单，独自）+ i + loqu（说）+ y →自言自语】

solitary adj. 孤独的；n. 隐士

solitude n. 孤独

195. solid 结实

consolidate v. 巩固；（使）坚强；合并

consolidation n. 合并，巩固

solidarity n. 团结，一致

solidify v. 巩固，（使）凝固，（使）团结

196. solv，solu 溶解，松开

absolve v. 赦免，免除

dissolute adj. 放荡的，无节制的

dissolve v. 使固体溶解

soluble adj. 可溶的；可以解决的

insoluble adj. 不溶解的；不能解决的

solvent adj. 有偿债能力的；n. 溶剂

insolvency n. 无力偿还；破产

197. son 声音

consonance n. 一致，调和；和音【con-（共同）+ son（声音）+ ance →一致，调和】

consonant adj. 调和的，一致的

dissonant adj. 不和谐的，不一致的【dis-（否定）+ son（声音）+ ant →不和谐的】

resonant adj.（声音）洪亮的；共鸣的

sonorous adj.（声音）洪亮的

ultrasonic adj. 超音速的，超声波的【ultra-（超）+ son（声音）+ ic →超音速的】

198. spec 看

circumspect adj. 慎重的【circum-（环绕）+ spec（看）+ t →四处看→慎重的】

inspection n. 检查，细看

inspector n. 检察员，检阅官

introspective adj. 自省的【intro-（向内）+ spec（看）+ tive →内省，自省】

introspection n. 反省，内省

perspective n.（判断事物的）角度，方法；透视法【per-（贯穿）+ spec（看）+ tive →看穿】

prospect v. 勘探；n. 期望；前景【pro-（前）+ spec（看）+ t →向前看→期望，前景】

respect v. 尊敬

retrospect n. 回顾；v. 回顾，回想

retrospective adj. 回顾的，追溯的【retro-（向后）+ spec（看）+ tive →向后看→回顾的】

spectacular adj. 壮观的，引人入胜的

suspect v. 怀疑；n. 嫌疑犯；adj. 可疑的【su-（=sub-，下）+ spec（看）+ t →在下面看，偷偷看→怀疑】

unsuspecting adj. 不怀疑的，无猜疑的，可信任的

199. spic 看

auspicious adj. 吉兆的【au + spic（看）+ ious →看到希望→吉兆的】

conspicuous adj. 显著的，显而易见的【con-（共同）+ spic（看）+ uous →都能看到的→显而易见的】

despicable adj. 可鄙的，卑劣的【de-（否定）+ spic（看）+ able →不值得看的→卑劣的】

despise v. 鄙视，蔑视

perspicacious adj. 独具慧眼的【per-（贯穿）+ spic（看）+ acious →看穿→独具慧眼的】

suspicion n. 怀疑，嫌疑【su-（= sub-，下）+ spic（看）+ ion →在下面看，偷偷看→怀疑】

suspicious adj. 怀疑的

200. spir 呼吸

aspire v. 向往，有志于【a + spir（呼吸）+ e →呼吸加快→向往某事物】

conspire v. 阴谋，共谋【con-（共同）+ spir（呼吸）+ e →同呼吸→共谋】

conspiracy n. 阴谋，共谋

expire v. 期满；去世【ex + spir（呼吸）+ e →呼吸完毕→去世】

inspire v. 吸入，吸（气）；鼓舞，激励；激起，唤起

inspired adj. 有创见的，有灵感的

inspiration n. 启示，灵感

uninspired adj. 无灵感的，枯燥的

201. struct 建造，结构

construct v. 建造，构成

constructive adj. 建设性的

destructible adj. 可破坏的

destruction n. 破坏，毁灭

indestructible adj. 不能破坏的，不可毁灭的

infrastructural adj. 基础结构的，基础设施的

infrastructure n. 基础设施，基础建设

obstruct v. 阻塞，截断

unobstructed adj. 没有阻碍的

reconstruct v. 重建，改造

restructure v. 重建，调整，重组

structure n. 结构；v. 建造

202. sum 总，概要

consummate adj. 完全的，完善的；v. 完成

summarily adv. 概括地；仓促地

summary n. 摘要，概要；adj. 摘要的，简略的

summation n. 总结，概要；总数，合计

203. surge 上升，起立

insurgent adj. 叛乱的，起事的；n. 叛乱分子

resurge v. 复活

resurgence n. 再起，复活，再现

surge v. 波涛汹涌，波动

upsurge n.（情绪）高涨

204. tach，tact 接触

attach v. 将某物附在（另一物）上

detach v. 分离，分遣【de-（否定）+ tach（接触）→分离】

detached adj. 分开的；超然的

contact v. 接触；互通信息

intact adj. 完整的，未动过的【in-（不）+ tact（接触）→未被接触过→完整的，未动过的】

tactile adj. 有触觉的

205. techn 技术，技巧

technique n. 技能，方法，手段

technocrat n. 技术管理人员【techn（技术）+ o + crat（统治）→统治技术的人】

206. tem 节制

abstemious adj. 有节制的，节俭的

intemperance n. 放纵，不节制，过度

temper v. 调和

temperate adj.（气候等）温和的；（欲望、饮食等）适度的，有节制的

207. tempor 时间，时代

contemporary adj. 同时代的；当代的；现代的【con-（共同）+ tempor（时间）+ ary →同时代的】

contemporaneous adj. 同时期的，同时代的

extemporaneous adj. 即席的，没有准备的【ex（外）+ tempor（时间）+ aneous →在时间安排之外的→即席的，没有准备的】

extemporize v. 即席演说

temporal adj. 时间的；世俗的

temporary adj. 暂时的，临时的

temporize v. 拖延；见风使舵

208. tend，tens，tent 拉

contend v. 竞争，争夺；据理力争，主张

contention n. 争论；论点

contentious adj. 好辩的，善争吵的

distend v.（使）膨胀，胀大

distent adj. 膨胀的；扩张的

extend v. 延展，延长；舒展（肢体）

extension n. 延伸，扩展

extensive adj. 广大的，广阔的；多方面的，广泛的

hypertension n. 高血压

tension n. 紧张，焦虑；张力

209. term 边界，结束

coterminous adj. 毗连的，有共同边界的

exterminate v. 消灭，灭绝

interminable adj. 无尽头的

terminal adj. 末端的；n. 终点站，终端

terminate v. 终止，结束

termination n. 终点

terminus n.（火车、汽车）终点站

210. the 神

apotheosis n. 神化；典范

atheism n. 无神论，不信神【a-（无）+ the（神）+ ism →无神论】

atheist n. 无神论者

pantheon n. 万神殿【pan-（全部）+ the（神）+ on →众神之地→万神殿】

theocracy n. 神权政治【the（神）+ o + cracy（统治）→神权政治】

211. tort 扭，弯曲

contort v.（使）扭曲；歪曲

distort v. 扭曲，弄歪

distortion n. 扭曲；曲解

undistorted adj. 未失真的

extort v. 勒索，敲诈【ex-（出）+ tort（扭）→勒索，敲诈】

extortion n. 敲诈，勒索

retort v. 反驳【re-（反）+ tort（扭）→反驳】

torment n. 折磨，痛苦

tornado n. 飓风，龙卷风

tortuous adj. 弯弯曲曲的

torture n. 酷刑，折磨；v. 对……施以酷刑

212. tox 毒，伤害

toxic adj. 有毒的，中毒的

toxin n. 毒素，毒质

intoxicate v. 使醉；使陶醉，使欣喜若狂

213. tract 拉，拖

detract v. 减去；贬低；转移【de-（否定）+ tract（拉）→减去；贬低】

detraction n. 贬低，诽谤

distract v. 分心，转移；使发狂【dis-（分开，离散）+ tract（拉）→分心，转移】

distracted adj. 心烦意乱的，精神不集中的

extract v. 拔出，强索【ex-（出）+ tract（拉）→拔出，强索】

protract v. 延长，拖长

retract v. 缩回，收回【re-（反）+ tract（拉）→缩回，收回】

subtract v. 减去，减掉

subtractive adj. 减法的；负的

tractable adj. 易于驾驭的，温顺的

tractability n. 温顺

intractable adj. 倔强的；难管的

214. tribut 给予，赠予

attribute n. 属性，品质；v. 把……归于

distribute v. 分发，分配某事物

redistribution n. 重新分配

tributary n. / adj. 支流（的）；进贡（的）

tribute n. 赞辞；贡物

215. tric 复杂，迷惑

extricable adj. 可解救的，能脱险的

extricate v. 摆脱，脱离；拯救，救出【ex-（出）+ tric（迷惑）+ ate →摆脱；拯救】

intricate adj. 复杂难懂的【in-（进）+ tric（迷惑）+ ate →复杂】

intricacy n. 错综，复杂，纷乱

216. trud，trus 伸

extrude v. 挤出，逐出；突出

intrude v.（把思想等）强加于；闯入

intruding adj. 入侵的，入侵性的

intrusion n. 侵扰，干扰

intrusively adv. 入侵地

obtrude v. 突出；强加

protrude v. 突出，伸出

unobtrusive adj. 不引人注目的

217. turb 扰乱

imperturbable adj. 冷静的，沉着的【im-（不）+ per + turb（扰乱）+ able →不被扰乱的】

turbulent adj. 导致动乱的，骚乱的

turmoil n. 混乱，骚乱

undisturbed adj. 未受干扰的，冷静的

218. vacu 空

evacuate v. 撤退；撤离

vacant adj. 空闲的，空缺的

vacuous adj. 发呆的；愚笨的

vacuum n. 真空

219. vad，vas 走

evade v. 逃避，规避

evasion n. 躲避，借口

evasive adj. 回避的，逃避的，托词的

invade v. 侵犯，侵入

invader n. 侵略者，侵犯者

invasion n. 侵略，入侵

invasive adj. 侵略性的，扩散的

pervade v. 弥漫，普及【per-（完全）+ vad（走）+ e →走遍→弥漫，普及】

pervasive adj. 弥漫的，遍布的

220. vag 流浪，漂泊

divagate v. 离题；漂泊

extravagance n. 奢侈，挥霍

vagabond n. 浪荡子，流浪者；adj. 流浪的

vagary n. 奇想，异想天开

vagrancy n. 游荡，流浪

vagrant adj. 漂泊的；n. 流浪汉，无赖

221. vari 变化

invariable adj. 恒定的，不变的

prevaricate v. 支吾其词，说谎【pre-（预先）+ vari（变化）+ cate →预先想好变化→说谎】

variable adj. 变化的，可变的

varied adj. 多变的，各种各样的

variability n. 变化性

variance n. 变化，变动；矛盾，不同

variegate v. 使……多样化

variegation n. 杂色，斑驳

variety n. 多样性；种类；变种

222. ven，vent 来

advent n. 到来，来临

circumvent v. 回避；用计谋战胜或规避【circum-（环绕）+ vent（来）→回避】

contravene v. 违背（法规、习俗等）【contra-（反）+ ven（来）+ e →违背】

convene v. 集合；召集【con-（共同）+ ven（来）+ e →集合】

intervene v. 干涉，介入【inter-（在……中间）+ ven（来）+ e →来到中间→干涉，介入】

intervention n. 干预，干涉，介入

223. ver 真实

aver v. 断言，确证

veracious adj. 诚实的，说真话的

veracity n. 真实性；诚实

verdict n. 判决，决定【ver（真实）+ dict（说）→根据事实说话→判决】

verifiable adj. 能作证的

verification n. 确认，查证

verified adj. 检验的，核实的

verify v. 证明，证实

verisimilitude n. 逼真【ver（真实）+ i + simil（像）+ itude →像真的→逼真】

verisimilar adj. 好像真实的；可能的

veritable adj. 确实的，名副其实的

224. verb 词语

proverb n. 谚语

verbal adj. 口头的；与言辞有关的

verbatim adj. 逐字的，照字面的

verbiage n. 啰嗦，冗长

verbose adj. 冗长的，啰嗦的

verbosity n. 冗长

225. verg 倾向

converge v. 会聚，集中于一点【con-（共同）+ verg（倾向）+ e →会聚】

convergent adj. 会聚的

diverge v. 分歧，分开【di-（二）+ verg（倾向）+ e →分开】

divergent adj. 分叉的，叉开的；发散的，扩散的；不同的

226. vert，vers 转

adversary n. 对手，敌手；adj. 对手的，敌对的【ad-（加强语气）+ vers（转）+ ary →转变→叛变】

adverse adj. 不利的；相反的；敌对的

inadvertence n. 漫不经心【in-（加强语气）+ ad-（加强语气）+ vert（转）+ ence →注意力转移】

inadvertent adj. 疏忽的；不经意的，无心的

inadvertently adv. 不小心地，疏忽地

controversial adj. 引起争论的【contro-（反）+ vers（转）+ ial →反着转→引起争论的】

uncontroversial adj. 未引起争论的

controversy n. 公开辩论，论战

controvert v. 反驳，驳斥

incontrovertible adj. 无可辩驳的

converse v. 谈话；adj. 逆向的；n. 相反的事物

conversely adv. 相反地，反过来

conversion n. 变换，转变

reconversion n. 再转变；恢复原状

convert v. 使改变（信仰等）；n. 改变信仰的人

convertible adj. 可转换的；n. 敞篷车

diverse adj. 不同的；多样的【di-（二，引申为多）+ vers（转）+ e →多样的】

diversify v.（使）多样化

diversity n. 多样，千变万化

divert v. 使事物转向；使娱乐

extrovert n. 性格外向者【extro-（外）+ vert（转）→性格外向者】

introvert n. 性格内向的人【intro-（内）+ vert（转）→性格内向者】

revert v. 恢复，回复到；重新考虑【re-（回，反）+ vert（转）→恢复】

reverse n. 反面；相反；v. 倒车；反转；adj. 反面的，颠倒的

reversal n. 反向，反转，倒转

reversible adj. 可逆的，可反转的

irreversible adj. 不能撤回的，不能取消的

reversion n. 返回（原状、旧习惯）；逆转

subvert v. 颠覆，推翻【sub-（下面）+ vert（转）→在下面转，在下面密谋→推翻，颠覆】

subversive adj. 颠覆性的，破坏性的

versatile adj. 多才多艺的；多用途的【vers（转）+ atile →样样玩得转→多才多艺的】

vertigo n. 眩晕【vert（转）+ igo →转晕→眩晕】

227. vi 道路

deviant adj. 越出常规的

deviate v. 越轨，脱离【de-（否定）+ vi（道路）+ ate →越轨】

deviation n. 背离

devious adj. 不正直的；弯曲的

viability n. 生存能力，存活力

viable adj. 可行的，能活下去的【vi（道路）+ able →有路可走的】

nonviable adj. 无法生存的

228. vigor 活力

invigorate v. 鼓舞，激励

invigorating adj. 使人有精神的，使人健壮的

vigorous adj. 精力旺盛的，健壮的

229. vis，vid 看，想

envisage v. 正视；想象

envision v. 想象，预想

improvise v. 即席而作【im-（不）+ pro-（前）+ vis（看）+ e →事前没有看过→即席而作】

prevision n. 先见，预感【pre-（前）+ vis（看）+ ion →预先看到→先见，预感】

provident adj. 深谋远虑的；节俭的【pro-（前）+ vid（看）+ ent →向前看的→深谋远虑的】

improvident adj. 不节俭的，无远见的

revise n. / v. 改变，修正【re-（再，又）+ vis（看）+ e →反复看→修正】

supervise v. 监督，管理【super-（上面）+ vis（看）+ e →从上面看→监督，管理】

visibility n. 能见度，可见性

vision n. 视觉，美景

visionary adj. 有远见的；幻想的；n. 空想家

vista n. 远景，街景

visual adj. 视觉的，看得见的

230. vit 活力，生命

revitalization n. 新生，复兴

revitalize v. 使重新充满活力

vital adj. 极其重要的；充满活力的

vitality n. 活力，精力；生命力

vitalize v. 激发活力

231. viv 活

convivial adj. 欢乐的，狂欢的

conviviality n. 欢乐；爱交际的性格

revival n. 苏醒；恢复；复兴

revive v. 使苏醒；再流行；使复活

survive v. 幸存【sur-（超）+ viv（活）+ e →超能活的→幸存】

vivacious adj. 活泼的，快活的

vivid adj. 清晰的，鲜艳的；大胆的；活泼的；逼真的

232. voc，vok 叫，喊

advocate v. 拥护，支持，鼓吹；n. 支持者，拥护者【ad-（加强语气）+ voc（喊）+ ate →摇旗呐喊→拥护，支持】

advocacy n. 拥护，支持

convoke v. 召集（会议等）【con-（一起）+ voc（喊）+ e →喊到一起→召集】

evocative adj. 唤起的，激起的

evoke v. 引起；唤起

invoke v. 祈求；恳求；（法律的）实施生效【in-（加强语气）+ vok（喊）+ e →祈求，恳求】

irrevocable adj. 无法取消的

provocation n. 挑衅，激怒

provocative adj. 挑衅的，煽动的

provoke v. 挑衅，激怒，煽动；引起【pro-（前）+ vok（喊）+ e →在……面前喊→挑衅】

unprovoked adj.（生气等）无缘无故的

revoke v. 撤销，废除；召回【re-（回，反）+ vok（喊）+ e →原来说的不算了→撤销，废除】

233. vol 意志，意愿

benevolent adj. 善心的，仁心的【bene-（好）+ vol（意愿）+ ent →善心的，仁心的】

malevolent adj. 有恶意的，恶毒的【mal-（坏）+ e + vol（意愿）+ ent →恶毒的】

involuntary adj. 无意的

volition n. 决断力，意志

voluntary adj. 自愿的，志愿的

234. volu，volv 转

convoluted adj. 旋绕的；费解的

evolve v. 使逐渐形成，进化【e + volv（转）+ e →向前转→进化】

evolution n. 发展，进化

involve v. 包含，含有；参与，牵涉【in-（进入）+ volv（转）+ e →包含，含有】

involvement n. 参与，连累

voluble adj. 健谈的；易旋转的

revolve v. 使旋转

235. vor 吃

carnivorous adj. 肉食动物的【carn（肉）+ i + vor（吃）+ ous →肉食动物的】

herbivorous adj. 食草的【herb（草）+ i + vor（吃）+ ous →食草的】

omnivorous adj. 杂食的；兴趣杂的【omni-（全部）+ vor（吃）+ ous →杂食的】

devour v. 吞食；如饥似渴地读

voracious adj. 狼吞虎咽的，贪婪的

voracity n. 贪婪

下　篇 经典文献导读

Unit 1

What Is Urban Planning?

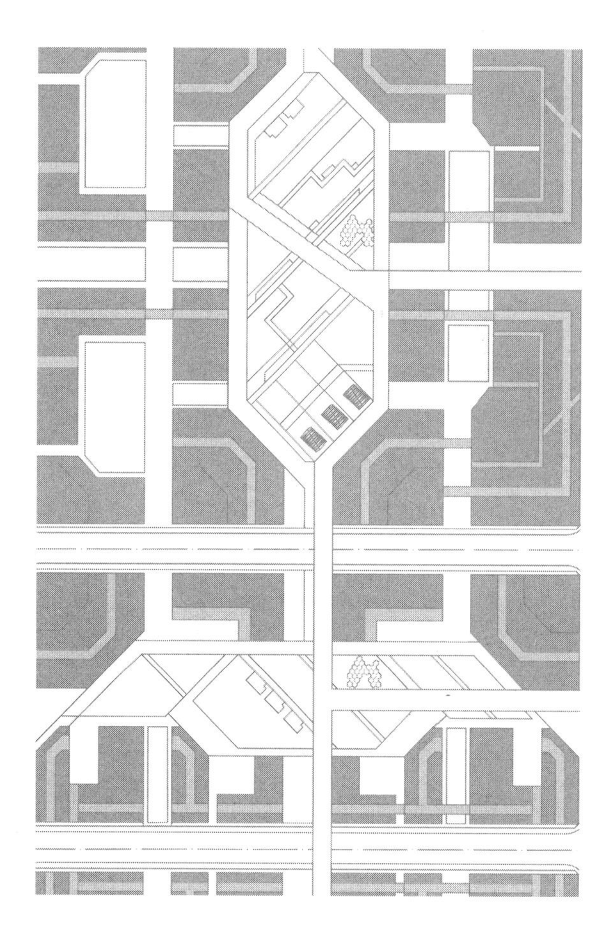

Urban planning is a technical and political process concerned with the use of land and the design of the urban environment. Urban planning is also referred to as urban and regional planning, regional planning, town planning, city planning, rural planning, urban development or some combination in various areas worldwide. It is considered an **interdisciplinary** field that includes social, engineering and design sciences. Urban planning is closely related to the field of urban design and some urban planners provide designs for streets, parks, buildings and other urban areas.

Urban planning guides orderly development in urban, suburban and rural areas. Although **predominantly** concerned with the planning of settlements and communities, urban planning is also responsible for the planning and development of water use and resources, rural and agricultural land, parks and **conserving** areas of natural environmental significance. Practitioners of urban planning are concerned with research and analysis, strategic thinking, architecture, urban design, public consultation, policy recommendations, implementation and management. Enforcement methodologies include governmental **zoning**, planning permissions, and building codes, as well as private **easements** and restrictive **covenants**.

Urban planners work with the **cognate** fields of architecture, landscape architecture, civil engineering, and public administration to achieve strategic, policy and **sustainability** goals. Early urban planners were often members of these cognate fields. Today urban planning is a separate, **independent** professional discipline. The discipline is the broader category that includes different sub-fields such as land-use planning, zoning, economic development, environmental planning, and transportation planning.

1 History

There is evidence of urban planning and designed communities dating back to the Mesopotamian, Indus Valley, Minoan, and Egyptian civilizations in the third **millennium** BCE. ***Archeologists** studying the ruins of cities in these areas find paved streets that were laid out at right angles in a **grid** pattern.* The idea of a planned out urban area **evolved** as different civilizations adopted it. Beginning in the 8th century BCE, Greek city states were primarily centered on **orthogonal** (or **grid-like**) plans. The ancient Romans, **inspired** by the Greeks, also used orthogonal plans for their cities. City planning in the Roman world was developed for military defense and public convenience. The spread of the Roman Empire **subsequently** spread the ideas of urban planning. As the Roman Empire declined, these ideas slowly disappeared. However, many cities in Europe still held onto the planned Roman city center. Cities in Europe from the 9th to 14th centuries, often grew **organically** and sometimes **chaotically**. But many hundreds of new towns were newly built according to **preconceived** plans, and many others were enlarged with newly planned **extensions**. Most of these were realized from the 12th to 14th centuries, with a peak-period at the end of the 13th. From the 15th century on, much more is recorded of urban design and the people that were **involved**. In this period, theoretical treatises on architecture and urban planning start to appear in which theoretical questions are addressed and designs of towns and cities are described and **depicted**. During the Enlightenment period, several European rulers **ambitiously** attempted to redesign capital cities. *During the Second French Republic, Baron Georges-Eugène Haussmann, under the direction of Napoleon Ⅲ, redesigned the city of Paris into a more modern capital, with long, straight, wide **boulevards**.*

Planning and architecture went through a **paradigm** shift at the turn of the 20th century. The industrialized cities of the 19th century grew at a **tremendous** rate. The pace and style of this industrial **construction** was largely **dictated** by the concerns of private business. The evils of urban life for the working poor were becoming increasingly evident as a matter for public concern. The **laissez-faire** style of government management of the economy, in fashion for most of the Victorian era, was starting to give way to a New **Liberalism** that **championed intervention** on the part of the poor and disadvantaged. Around 1900, theorists began developing urban planning models to **mitigate** the consequences of the industrial age, by providing citizens, especially factory workers, with healthier environments.

Urban planning started to become professionalized during this time. The Town and Country Planning Association was founded in 1899 and the first academic course in Great Britain on urban planning was offered by the University of Liverpool in 1909. In the 1920s, the ideas of **modernism** and **uniformity** began to surface in urban planning, and lasted until the 1970s. Many planners started to believe that the ideas of modernism in urban planning led to higher crime rates and social problems. Urban planners now focus more on individualism and **diversity** in urban centers.

2 Theories

Planning theory is the body of scientific concepts, definitions, behavioral relationships, and assumptions that define the body of knowledge of urban planning. There are ten procedural theories of planning that remain the principal theories of planning procedure today: the **blueprint** planning, the **synoptic** planning, the **participatory** planning, the **incremental** planning, the mixed **scanning** model, the **transactive** planning, the **advocacy** planning, the radical planning, the **bargaining** model, and the **communicative** planning.

3 Technical aspects

Urban planning includes techniques such as: **predicting** population growth, zoning, geographic mapping and analysis, analyzing park space, surveying the water supply, identifying transportation patterns, recognizing food supply demands, **allocating** healthcare and social services, and analyzing the impact of land use.

In order to predict how cities will develop and estimate the effects of their interventions, planners use various models. These models can be used to indicate relationships and patterns in **demographic,** geographic, and economic data. They might deal with short–term issues such as how people move through cities, or long–term issues such as land use and growth.

Building codes and other regulations **dovetail** with urban planning by governing how cities are constructed and used from the individual level.

4 Urban planners

An urban planner is a professional who works in the field of urban planning for the purpose of **optimizing** the effectiveness of a community's land use and infrastructure. They formulate plans for the development and management of urban and suburban areas, typically analyzing land use **compatibility** as well as economic, environmental and social trends. *In developing the plan for a community (whether commercial, residential, agricultural, natural or recreational), urban planners must also consider a wide array of issues such as* *sustainability, air pollution, traffic* ***congestion****, crime, land values, legislation and zoning codes.*

The importance of the urban planner is increasing throughout the 21st century, as modern society begins to face issues of increased population growth, climate change and **unsustainable** development. An urban planner could be considered a green collar professional.

Some researchers suggest that urban planners around the world work in different "planning cultures", adapted to their local cities and cultures. However, there are several basic knowledge, skills and abilities that are **universalizable** to the urban planning profession even amongst planners in different countries and regions.

Notes

[1] 本文源自维基百科（From Wikipedia, the free encyclopedia, https://www.wikipedia.org/）。

[2] 关于"**1 History**"，详见"**Unit 2 History of Urban Planning (1)**"和"**Unit 3 History of Urban Planning (2)**"。

[3] 关于"**2 Theories**"，详见"**Unit 7 Theories of Urban Planning**"。其他问题：the incremental planning，渐进性规划；the mixed scanning model，混合审视模式；the transactive planning, 协商性规划；the advocacy planning，倡导性规划；the communicative planning，联络性或沟通性规划。

[4] 本文主要相关文献：孙施文，1987；于泓，2000；张庭伟，1999；Davidoff，1965；Etzioni，1967；Hall，2002；Innes，1995；Lindblom，1959；Sager，1994；Taylor，1998。

[5] 课文中下划线部分为有关城市规划理论的重要论述，需强化阅读，每篇课文习题部分均要求将下划线句子或段落译成中文。

[6] 课文中下划线斜体字部分为复杂长句，其解析详见下文"**Complex Sentences**"。

New Words

advocacy n. 拥护，支持〖前缀 3. ad-; 词根 232. voc〗

allocate v. 分配，分派

ambitiously adv. 雄心勃勃地〖前缀 5. ambi-〗

archeologist n. 考古学家〖前缀 8. arch-〗

bargain v. 讨价还价，做交易

BCE: Before the Common Era 公元前

blueprint n. 蓝图

boulevard n. 大马路，林荫大道

champion v. 捍卫，声援

chaotically adv. 混乱地

cognate adj. 同宗的，同根的

communicative adj. 爱说话的，好社交的

compatibility n. 和睦相处，适合，通用性〖前缀 16. col–; 词根 153. pati〗

congestion n. 拥挤，堵塞

conserve v. 保护，保存，保藏〖前缀 16. con–; 词根 188. serv〗

construction n. 建造，建筑物〖前缀 16. con–; 词根 201. struct〗

covenant n. 协议，协定

demographic adj. 人口统计的，人口统计学的〖词根 54. dem; 词根 84. graph(y)〗

depict v. 描绘，描画，描述〖前缀 18. de–; 词根 165. pict〗

dictate v. 命令，指示〖词根 55. dict〗

diversity n. 多样性〖前缀 19. di–; 词根 226. vers〗

dovetail n. 榫头；v. 吻合

easement n. 地役权；附属建筑物

extension n. 伸展，扩大，延长〖前缀 26. ex–; 词根 208. tens〗

grid n. 格子，方格

incremental adj. 增加的〖前缀 34. in–; 词根 45. cre〗

independent adj. 自主的，独立的〖前缀 34. in–〗

inspire v. 鼓舞，激励，启迪〖前缀 34. in–; 词根 200. spir〗

interdisciplinary adj. 跨学科的〖前缀 35. inter–〗

intervention n. 介入，干涉，干预〖前缀 35. inter–; 词根 222. vent〗

involve v. 包含，牵扯，使参与〖前缀 34. in–; 词根 234. volv〗

laissez–faire adj. 放任主义的，自由放任的

liberalism n. 自由主义〖词根 108. liber〗

millennium n. 一千年，千禧年〖前缀 mill–, 千；词根 11. enn〗

mitigate v. 使缓和，使减轻

modernism n. 现代主义

optimize v. 使最优化

organically adv. 有机地，自然发展地

orthogonal adj. 直角的，互相垂直的〖词根 149. ortho〗

paradigm n. 范例，样式

participatory adj. 提供参加机会的，供人分享的

preconceived adj. 预想的，先入为主的〖前缀 57. pre–〗

predict v. 预言，预测，预示〖前缀 57. pre–; 词根 55. dict〗

predominantly adv. 显著地，占优势地

scan v. 扫描；细看，细查

sustainability n. 可持续性

synoptic adj. 提纲的，概要的

transact v. 办理，交易；商议，谈判〖前缀 71. trans-; 词根 2. act〗

tremendous adj. 极大的，巨大的

uniformity n. 均匀性，单调，无变化〖前缀 76. uni-; 词根 76. form〗

universal adj. 普遍的，通用的，全世界的〖前缀 76. uni-〗

universalizable adj. 可通用的，可普遍应用的〖前缀 76. uni-〗

unsustainable adj. 不可持续的

zoning n. 区划，分地带

Complex Sentences

[1] **Archeologists** studying the ruins of cities in these areas **find paved streets** that were laid out at right angles in a grid pattern.

黑体字部分为句子的主干（主语、谓语和宾语），studying 引导的现在分词短语作定语，修饰前面的 archaeologists, that 引导的从句为定语从句，修饰前面的 paved streets。

[2] During the Second French Republic, **Baron Georges-Eugène Haussmann**, under the direction of Napoleon Ⅲ, **redesigned the city of Paris into a more modern capital**, with long, straight, wide boulevards.

黑体字部分为句子的主干（主语、谓语和宾语），during 引导的介词短语和 under 引导的介词短语均作状语，with 引导的介词短语作定语，修饰前面的 capital。

[3] In developing the plan for a community (whether commercial, residential, agricultural, natural or recreational), **urban planners must also consider a wide array of issues** such as sustainability, air pollution, traffic congestion, crime, land values, legislation and zoning codes.

黑体字部分为句子的主干（主语、谓语和宾语），in 引导的介词短语作状语，其中括号内容为解释性插入语，such as 引导的介词短语作定语，修饰前面的 issues。

Exercises

[1] 将课文中下划线句子或段落译成中文。

[2] 读熟或背诵上述复杂长句。

[3] 解析下列单词［解析单词是指根据单词中包含的词根、前缀推导出单词的含义，例如，apathy（n. 漠然，冷淡），解析式为 :a-（无）+ path（感情）+ y →无感情→冷漠，冷淡］。解析的单词选

自上文“**New Words**”。

advocacy n. 拥护，支持	involve v. 包含，牵扯，使参与
compatibility n. 和睦相处，适合，通用性	millennium n. 一千年，千禧年
demographic adj. 人口统计（学）的	predict v. 预言，预测，预示
extension n. 伸展，扩大，延长	transact v. 办理，交易；商议，谈判
intervention n. 介入，干涉，干预	uniformity n. 均匀性，单调，无变化

Unit 2

History of Urban Planning (1)

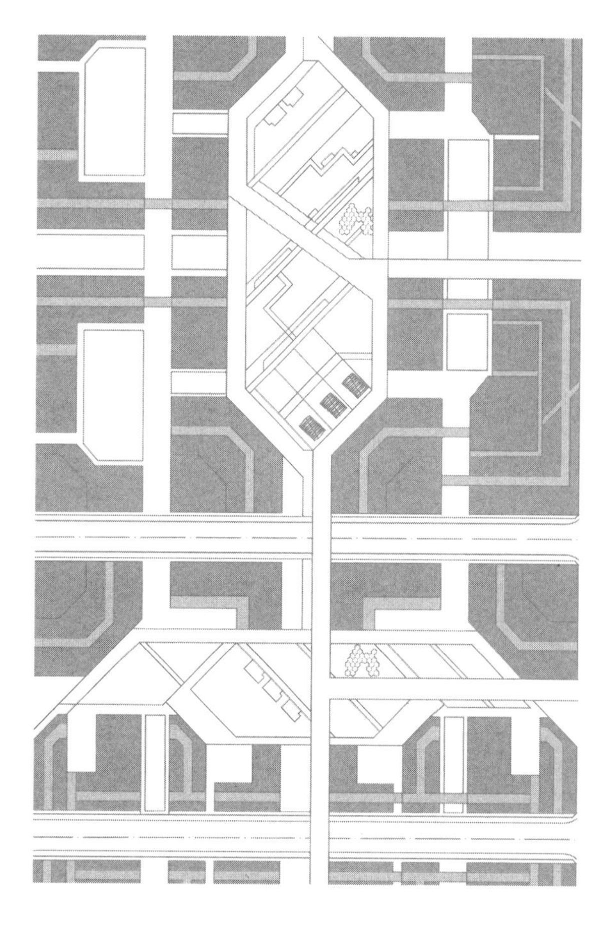

This article **delineates** the history of urban planning, a technical and political process concerned with the use of land and design of the urban environment. The history of urban planning runs **parallel** to the history of the city, as planning is in evidence at some of the earliest known urban sites.

1 Pre-classical

The pre-Classical and **Classical** periods saw a number of cities laid out according to fixed plans, though many tended to develop **organically**. Designed cities were characteristic of the Minoan, Mesopotamian, Harrapan, and Egyptian civilizations of the third **millennium** BC. The first recorded description of urban planning appears in the Epic of Gilgamesh.

Distinct characteristics of urban planning from remains of the cities of Harappa, Lothal, Dholavira, and Mohenjo-daro in the Indus Valley Civilization (in modern-day northwestern India and Pakistan) lead **archeologists** to interpret them as the earliest known examples of **deliberately** planned and managed cities. The streets of many of these early cities were paved and laid out at right angles in a **grid** pattern, with a **hierarchy** of streets from major **boulevards** to residential **alleys**. **Archaeological** evidence suggests that many Harrapan houses were laid out to protect from noise and to **enhance** residential privacy;many also had their own water wells, probably both for **sanitary** and for ritual purposes. These ancient cities were **unique** in that they often had drainage systems, seemingly tied to a well-developed ideal of urban **sanitation**. Cities laid out on the grid plan could have been an **outgrowth** of agriculture based on **rectangular** fields. Many Central American civilizations also planned their cities, including **sewage** systems and running water. In Mexico, Tenochtitlan, built on an island in Lake Texcoco in the present-day the Federal District in central Mexico, served as the capital of the Aztec empire. At its height, Tenochtitlan was one of the largest cities in the world, with over 200, 000 inhabitants.

2 Greco-Roman empires

Traditionally, the Greek philosopher Hippodamus (5th century BC) is regarded as the first town planner and "inventor" of the **orthogonal** urban layout. Aristotle called him "the father of city planning", and until well into the 20th century, he was indeed regarded as such. This is, however, only partly justified. The Hippodamian plan that was called after him, is an orthogonal urban layout with more or less square street blocks. Archaeological finds from ancient Egypt—among others—demonstrate that Hippodamus cannot truly have been the inventor of this layout. Aristotle's **critique** and indeed **ridicule** of Hippodamus is perhaps the first known example of a **criticism** of urban planning.

From about the late 8th century BC on, Greek city-states started to found colonies along the coasts of the Mediterranean, which were centered on newly created towns and cities with more or less regular orthogonal plans. Gradually, the new layouts became more regular. After the city of Miletus was destroyed by the Persians in 494 BC, it was rebuilt in a regular form that, according to tradition, was determined by the ideas of Hippodamus. Regular orthogonal plans particularly appear to have been laid out for new colonial cities and cities that were rebuilt in a short period of time after **destruction**.

Following in the tradition of Hippodamus about a century later, Alexander ***commissioned*** *the architect Dinocrates to lay out his new city of Alexandria, the grandest example of idealized urban planning of the ancient* ***Hellenistic*** *world, where the city's regularity was* ***facilitated*** *by its level site near a mouth of the Nile.*

The ancient Romans also employed regular orthogonal structures on which they molded their colonies. They probably were **inspired** by Greek and Hellenic examples, as well as by regularly planned cities that were built by the Etruscans in Italy. The Roman engineer Vitruvius established principles of good design whose influence is still felt today.

The Romans used a **consolidated** scheme for city planning, developed for civil convenience. The basic plan consisted of a central **forum** with city services, surrounded by a **compact**, **rectilinear** grid of streets. A river sometimes flowed near or through the city, providing water, transport, and sewage **disposal**. Hundreds of towns and cities were built by the Romans throughout their empire. Many European towns, such as Turin, **preserve** the remains of these schemes, which show the very logical way the Romans designed their cities. They would lay out the streets at right angles, in the form of a square grid. All roads were equal in width and length, except for two, which were slightly wider than the others. The decumanus, running east–west, and the cardo, running north–south, **intersected** in the middle to form the centre of the grid. All roads were made of carefully fitted flag stones and filled in with smaller, hard–packed rocks and pebbles. Bridges were constructed where needed. Each square marked by four roads was called an insula, the Roman **equivalent** of a modern city block.

Each insula was about 80 yards (73 m) square. As the city developed, it could eventually be filled with buildings of various shapes and sizes and **criss–crossed** with back roads and alleys.

The city may have been surrounded by a wall to protect it from **invaders** and to mark the city limits. Areas outside city limits were left open as farmland. At the end of each main road was a large gateway with watchtowers. A **portcullis** covered the opening when the city was under siege, and additional watchtowers were constructed along the city walls. An **aqueduct** was built outside the city walls.

The development of Greek and Roman urbanization is relatively well–known, as there are relatively many written sources, and there has been much attention to the subject since the Romans and Greeks are generally regarded as the main **ancestors** of modern Western culture. It should not be forgotten, though, that there were also other cultures with more or less urban settlements in Europe, primarily of Celtic origin.

3 Medieval Europe

After the gradual ***disintegration*** *and fall of the West–Roman empire in the 5th century and the devastation by the* ***invasions*** *of Huns, Germanic peoples, Byzantines, Moors, Magyars, and Normans in the next five centuries, little remained of urban culture in western and central Europe.* In the 10th and 11th centuries, though, there appears to have been a general improvement in the political stability and economy. This made it possible for trade and craft to grow and for the monetary economy and urban culture to **revive**. **Initially**, urban culture recovered particularly in existing settlements, often in remnants of Roman towns and cities, but later on, ever more towns were created anew. Meanwhile, the population of western Europe increased rapidly and the utilized agricultural area grew with it. The agricultural areas of existing villages were **extended** and new villages and towns were created in uncultivated areas as cores for new **reclamations**.

Urban development in the early Middle Ages, characteristically focused on a **fortress**, a fortified **abbey**, or a (sometimes **abandoned**) Roman nucleus, occurred "like the **annular** rings of a tree", whether in an extended village or the centre of a larger city. Since the new centre was often on high, defensible ground, the city plan took on an **organic** character, following the irregularities of **elevation contours** like the shapes that result from agricultural **terracing**.

In the 9th to 14th centuries, many hundreds of new towns were built in Europe, and many others were enlarged with newly planned extensions. These new towns and town extensions have played a very important role in the shaping of Europe's geographical structures as they in modern times. New towns were founded in different parts of Europe from about the 9th century on, but most of them were realized from the 12th to 14th centuries, with a peak-period at the end of the 13th. All kinds of landlords, from the highest to the lowest rank, tried to found new towns on their estates, in order to gain economical, political or military power. The settlers of the new towns generally were attracted by fiscal, economic, and **juridical** advantages granted by the founding lord, or were forced to move from elsewhere from his estates. Most of the new towns were to remain rather small (as for instance the **bastides** of southwestern France), but some of them became important cities, such as Cardiff, Leeds, 's-Hertogenbosch, Montauban, Bilbao, Malmö, Lübeck, Munich, Berlin, Bern, Klagenfurt, Alessandria, Warsaw and Sarajevo.

From the evidence of the **preserved** towns, it appears that the formal structure of many of these towns was **willfully** planned. The newly founded towns often show a marked regularity in their plan form, in the sense that the streets are often straight and laid out at right angles to one another, and that the house lots are rectangular, and originally largely of the same size. One very clear and relatively extreme example is Elburg in the Netherlands, dating from the end of the 14th century. Looking at town plans such as the one of Elburg, it clearly appears that it is impossible to maintain that the straight street and the **symmetrical,** orthogonal town plan were new inventions from the **Renaissance**, and, therefore, typical of modern times.

The deep **depression** around the middle of the 14th century marked the end of the period of great urban expansion. Only in the parts of Europe where the process of urbanization had started relatively late, as in eastern Europe, was it still to go on for one or two more centuries. It would not be until the Industrial Revolution that the same level of expansion of urban population would be reached again, although the number of newly created settlements would remain much lower than in the 12th and 13th centuries.

4 Renaissance Europe

Florence was an early model of the new urban planning, which took on a **star-shaped** layout adapted from the new star **fort**, designed to resist **cannon** fire. This model was widely imitated, reflecting the enormous cultural power of Florence in this age; "the Renaissance was **hypnotised** by one city type which for a century and a half—from Filarete to Scamozzi—was impressed upon **utopian** schemes: this is the star-shaped city". **Radial** streets extend outward from a defined centre of military, communal or spiritual power.

Only in ideal cities did a centrally planned structure stand at the heart, as in Raphael's Sposalizio of 1504. As built, the unique example of a rationally planned **quattrocento** new city centre, that of Vigevano (1493–1495), **resembles** a closed space instead, surrounded by **arcading**.

Filarete's ideal city, building on Leon Battista Alberti's De re aedificatoria, was named "Sforzinda" in **compliment** to his **patron**;its twelve-pointed shape, **circumscribable** by a "perfect" Pythagorean figure, the circle, took no heed of its **undulating** terrain in Filarete's **manuscript**. This process occurred in cities, but ordinarily not in the industrial suburbs characteristic of this era, which remained disorderly and characterized by crowding and organic growth.

Following the 1695 **bombardment** of Brussels by the French troops of King Louis XIV, in which a large part of the city centre was destroyed, Governor Max Emanuel proposed using the **reconstruction** to completely change the layout and architectural style of the city. His plan was to **transform** the medieval city into a city of the new **baroque** style, modeled on Turin, with a logical street layout, with straight avenues offering long, **uninterrupted** views **flanked** by buildings of a uniform size. This plan was opposed by residents and **municipal** authorities, who wanted a rapid reconstruction, did not have the resources for **grandiose** proposals, and **resented** what they considered the **imposition** of a new, foreign, architectural style. In the actual reconstruction, the general layout of the city was **conserved**, but it was not identical to that before the **cataclysm**. Despite the necessity of rapid reconstruction and the lack of financial means, authorities did take several measures to improve traffic flow, sanitation, and the aesthetics of the city. Many streets were made as wide as possible to improve traffic flow.

5 Enlightenment Europe and America

During this period, rulers often **embarked** on **ambitious** attempts at redesigning their capital cities as a showpiece for the **grandeur** of the nation. Disasters were often a major **catalyst** for planned reconstruction. *An exception to this was in London after the Great Fire of 1666 when, despite many radical rebuilding schemes from architects such as John Evelyn and Christopher Wren, no large-scale redesigning was achieved due to the complexities of rival ownership claims.* However, improvements were made in hygiene and fire safety with wider streets, stone construction and **access** to the river.

The Great Fire did, however, **stimulate** thinking about urban design that influenced city planning in North America. The Grand Model for the Province of Carolina, developed in the aftermath of the Great Fire, established a **template** for colonial planning. The famous Oglethorpe Plan for Savannah (1733) was in part influenced by the Grand Model.

In contrast, after the 1755 Lisbon earthquake, King Joseph I of Portugal and his ministers immediately launched efforts to rebuild the city. The architect Manuel da Maia boldly proposed **razing** entire sections of the city and "laying out new streets without restraint". This last **option** was chosen by the king and his minister. Keen to have a new and perfectly ordered city, the king commissioned the construction of big squares, rectilinear, large **avenues** and widened streets—the new **mottos** of Lisbon. The Pombaline buildings were among the earliest **seismically** protected constructions in Europe.

An even more ambitious reconstruction was carried out in Paris. In 1852, Baron Georges-Eugène Haussmann was commissioned to remodel the Medieval street plan of the city by demolishing **swathes** of the old quarters and laying out wide boulevards, extending outwards beyond the old city limits. Haussmann's project **encompassed** all aspects of urban planning, both in the centre of Paris and in the surrounding districts, with regulations **imposed** on building **facades**, public parks, **sewers** and water works, city facilities, and public monuments. Beyond aesthetic and sanitary considerations, the wide **thoroughfares** facilitated troop movement and policing.

A **concurrent** plan to extend Barcelona was based on a scientific analysis of the city and its modern requirements. It was drawn up by the Catalan engineer Ildefons Cerdà to fill the space beyond the city walls after they were demolished from 1854. He is **credited** with inventing the term "urbanization" and his approach was **codified** in his General Theory of Urbanization (1867). Cerdà's Eixample (Catalan for "extension") consisted of 550 regular blocks with **chamfered** corners to facilitate the movement of **trams**, crossed by three wider avenues. His objectives were to improve the health of the inhabitants, towards which the blocks were built around central gardens and **orientated** NW-SE to maximize the sunlight they received, and assist social **integration**.

Notes

[1] 本文源自维基百科（From Wikipedia, the free encyclopedia, https://www.wikipedia.org/）。

[2] 本文主要相关文献：李德华，2001；沈玉麟，1989；吴志强，等，2010；张京祥，2005。

[3] 课文中下划线部分为有关城市规划理论的重要论述，需强化阅读，每篇课文习题部分均要求将下划线句子或段落译成中文。

[4] 课文中下划线斜体字部分为复杂长句，其解析详见下文"**Complex Sentences**"。

New Words

abandon v. 放弃，放纵〖前缀 1. a-；词根 18. band〗

abbey n. 修道院，大教堂

access n. 接近，进入，通路；v. 接近，使用〖前缀 3. ac-；词根 30. cess〗

alley n. 胡同，小巷，小径

ambitious adj. 有雄心的，有野心的〖前缀 5. ambi-〗

ancestor n. 祖先，祖宗

annular adj. 环的，环形的

aqueduct n. 引水渠，沟渠，高架渠〖词根 13. aqu；词根 56. duct〗

arcade n. 拱廊，有拱廊的街道

archeological adj. 考古学的〖前缀 8. arch-〗

archeologist n. 考古学家〖前缀 8. arch-〗

archeology n. 考古学〖前缀 8. arch-〗

avenue n. 林荫路，大街；途径

baroque adj. 巴洛克风格的，过分装饰的

bastide n. 中世纪专为防御而建的乡村或城镇

bombardment n. 炮击，轰炸

boulevard n. 大马路，林荫大道

canon n. 大炮

cataclysm n.（突然降临的）大灾难

catalyst n. 催化剂

chamfer v. 斜切

circumscribe v. 限制，限定；在……周围画线〖前缀 15. circum-；词根 181. scribe〗

classical adj. 古典的

codify v. 变成法典

commission n. 委员会，委员；v. 委任，授予

compact adj. 紧凑的，简洁的

compliment n. 恭维，敬意

concurrent adj. 同时发生的，同时完成的〖前缀 16. con-; 词根 52. cur〗

conserve v. 保护，保存，保藏〖前缀 16. con-; 词根 188. serv〗

consolidate v. 巩固，加强〖前缀 16. con-; 词根 195. solid〗

contours n. 等高线

credit v. 相信，信任〖词根 46. cred〗

criss-cross adj. 纵横交错的，交叉的

criticism n. 评论，批评，批判〖词根 48. crit〗

critique v. & n. 批判，发表评论〖词根 48. crit〗

deliberately adv. 深思熟虑地，故意地〖前缀 18. de-; 词根 108. liber〗

delineate v. 勾画，描述

depression n. 沮丧，萎靡不振〖前缀 18. de-; 词根 172. press〗

destruction n. 破坏，毁灭〖前缀 18. de-; 词根 201. struct〗

disintegration n. 瓦解，崩溃

disposal n. 处置，处理〖前缀 20. dis-; 词根 170. pos〗

elevation n. 高地，高度，海拔〖词根 106. lev〗

embark v. 着手，从事

encompass v. 包含，包括；围绕，包围〖前缀 24. en-〗

enhance v. 提高，加强

enlightenment n. 启迪，启发，教化

equivalent n. 对等物；adj. 相等的，相交的〖词根 58. equ〗

extend v. 延伸，扩大，推广〖前缀 26. ex-; 词根 208. tend〗

facade n.（建筑物的）正面，门面，立面〖词根 62. face〗

facilitate v. 促进，助长〖词根 61. fac〗

flank v. 位于……的侧面

fort n. 堡垒，要塞

fortress n. 堡垒，要塞

forum n. 论坛，讨论会；集会的公共场所

grandeur n. 宏伟，伟大〖词根 83. grand〗

grandiose adj. 宏伟的；浮夸的〖词根 83. grand〗

grid n. 格子，方格

Hellenistic adj. 希腊风格的，希腊文化的

hierarchy 层次，等级，等级制度〖词根 14. arch(y)〗

hypnotize v. 使着迷，对……施催眠术

impose v. 强加〖前缀 33. im–; 词根 170. pos〗

imposition n. 强加，被迫接受〖前缀 33. im–; 词根 170. pos〗

initially adv. 开始，最初

inspire v. 鼓舞，激励〖前缀 34. in–; 词根 200. spir〗

integrate v. 使成整体

intersect v. 相交，交叉〖前缀 35. inter–; 词根 182. sect〗

invader n. 侵略者〖前缀 34. in–; 词根 219. vad〗

invasion n. 侵略，入侵〖前缀 34. in–; 词根 219. vas〗

juridical adj. 裁判的，司法的

manuscript n. 手稿，底稿，原稿〖词根 119. manu; 词根 181. script〗

millennium n. 一千年，千年期〖前缀 mill–, 千；词根 11. enn〗

motto n. 座右铭，格言

municipal adj. 市的，市政的

option n. 选择

organic adj. 有机的，自然发展的

organically adv. 有机地，自然发展地

orient v. 标定方向，使……向东方；adj. 东方的，新生的；n. 东方〖词根 148. ori〗

orientate v. = orient〖词根 148. ori〗

orthogonal adj. 直角的，直交的，矩形的〖词根 149. ortho〗

outgrowth n. 自然结果〖前缀 47. out–〗

parallel adj. 平行的，类似的；v. 平行于；n. 平行线

patron n. 赞助人，资助人，(宗) 守护神〖词根 154. patri〗

portcullis n. 吊闸，铁闸

preserve v. 保护，保持，保存〖词根 188. serv〗

quattrocento n. 公元 15 世纪 (文艺复兴初期)

radial adj. 辐射状的，放射式的

raze v. 彻底摧毁，将……夷为平地

reclamation n. 开垦；改造〖前缀 61. re–; 词根 38. clam〗

reconstruction n. 重建，再现〖前缀 61. re–; 词根 201. struct〗

rectangular adj. 矩形的，成直角的〖词根 177. rect〗

rectilinear adj. 直线的〖词根 177. rect〗

renaissance n. 文艺复兴，文艺复兴时期

resemble v. 相像，类似于〖词根 184. semble〗

resent v. 怨恨，愤恨，厌恶〖前缀 61. re–; 词根 186. sent〗

revive v. 复活，恢复〖前缀 61. re–; 词根 231. viv〗

ridicule n. 嘲笑，奚落

sanitary adj. 清洁的，卫生的

sanitation n. 卫生系统或设备

seismic adj. 地震的，地震引起的

sewage n. 污物，污水

sewer n. 污水管，下水道

star–shaped adj. 星形的

stimulate 刺激，激励

swath n. 长而宽的一条，长而宽的地带

symmetrical adj. 对称的，匀称的〖前缀 70. sym–; 词根 126. metr〗

template n. 样板，模板

terrace n. 台阶，阶地

thoroughfare n. 大道，大街

tram n. 有轨电车

transform v. 改变，改观〖前缀 71. trans–; 词根 76. form〗

undulate v. 起伏，波动

uninterrupted adj. 不间断的，连续的〖前缀 76. un–; 前缀 35. inter–; 词根 178. rupt〗

unique adj. 唯一的，独一无二的〖前缀 76. uni–〗

utopian adj. 乌托邦的，空想的，不切实际的

willfully adj. 任性的，固执的；故意的，存心的

Complex Sentences

[1] Following in the tradition of Hippodamus about a century later, **Alexander commissioned the architect Dinocrates to lay out his new city of Alexandria**, the grandest example of idealized urban planning of the ancient Hellenistic world, where the city's regularity was facilitated by its level site near a mouth of the Nile.

黑体字部分为句子的主干，following 引导的现在分词短语作状语，the grandest example of idealized urban planning of the ancient Hellenistic world 为前面 his new city of Alexandria 的同位语，where 引导的从句为定语从句，修饰前面的 his new city of Alexandria。

[2] After the gradual disintegration and fall of the West–Roman empire in the 5th century and the devastation by the invasions of Huns, Germanic peoples, Byzantines, Moors, Magyars, and Normans

in the next five centuries, **little remained of urban culture in western and central Europe**.

黑体字部分为句子的主干，系倒装结构，正常语序为 little of urban culture in western and central Europe remained。After 引导的介词短语很长，作状语。

[3] **An exception to this was in London** after the Great Fire of 1666 when, despite many radical rebuilding schemes from architects such as John Evelyn and Christopher Wren, no large-scale redesigning was achieved due to the complexities of rival ownership claims.

黑体字部分为句子的主干，after the Great Fire of 1666 作时间状语，when 引导的从句为时间状语从句，其中 despite 引导的介词短语将 when 引导的时间状语从句隔开，增加了阅读的难度。

Exercises

[1] 将课文中下划线句子或段落译成中文。

[2] 读熟或背诵上述复杂长句。

[3] 解析下列单词 [解析单词是指根据单词中包含的词根、前缀推导出单词的含义，例如，apathy（n. 漠然，冷淡），解析式为：a-（无）+ path（感情）+ y →无感情→冷漠，冷淡]。解析的单词选自上文"**New Words**"。

abandon v. 放弃，放纵	impose v. 强加
aqueduct n. 引水渠，沟渠，高架渠	intersect v. 相交，交叉
circumscribe v. 限制，限定；在……周围画线	invader n. 侵略者
concurrent adj. 同时发生的，同时完成的	manuscript n. 手稿，底稿，原稿
consolidate v. 巩固，加强	reclamation n. 开垦；改造
destruction n. 破坏，毁灭	resent v. 怨恨，愤恨，厌恶
disposal n. 处置，处理	revive v. 复活，恢复
elevation n. 高地，高度，海拔	symmetrical adj. 对称的，匀称的
equivalent n. 对等物；adj. 相等的，相交的	transform v. 改变，改观
hierarchy 层次，等级，等级制度	uninterrupted adj. 不间断的，连续的

Unit 3

History of Urban Planning (2)

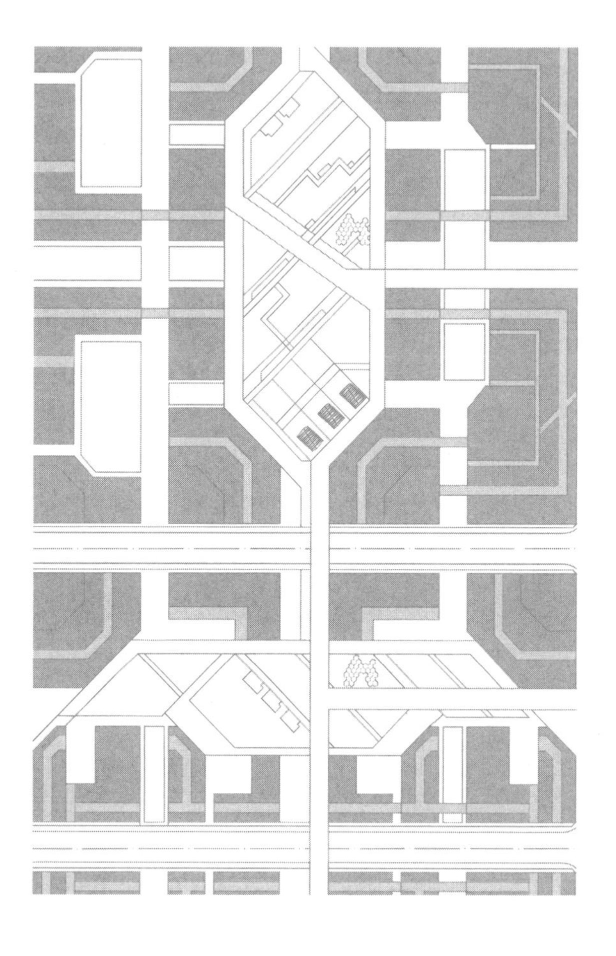

6 Modern urban planning

Planning and architecture went through a **paradigm** shift at the turn of the 20th century. The industrialized cities of the 19th century had grown at a **tremendous** rate, with the pace and style of building largely **dictated** by private business concerns. The **evils** of urban life for the working poor were becoming increasingly evident as a matter for public concern. *The **laissez-faire** style of government management of the economy, in fashion for most of the Victorian era, was starting to give way to a New **Liberalism** that **championed intervention** on the part of the poor and **disadvantaged**.* Around 1900, theorists began developing urban planning models to **mitigate** the consequences of the industrial age, by providing citizens, especially factory workers, with healthier environments.

Modern **zoning**, which enabled planners to legally **demarcate** sections of cities for different functions, **originated** in Prussia, and spread to Britain, the USA, and Scandinavia. Public health was cited as a **rationale** for keeping cities organized.

6.1 Garden city movement

The first major urban planning theorist was Sir Ebenezer Howard, who **initiated** the garden city movement in 1898. This was inspired by earlier planned communities built by industrial **philanthropists** in the countryside, such as Cadburys' Bournville, Lever's Port Sunlight and George Pullman's **eponymous** Pullman in Chicago. All these settlements **decentralized** the working environment from the centre of the cities, and provided a healthy living space for the factory workers. Howard generalized this achievement into a planned movement for the country as a whole. *He was also influenced by the work of economist Alfred Marshall who argued in 1884 that industry needed a supply of labor that could in theory be supplied anywhere, and that companies have an* ***incentive*** *to improve workers living standards as the company bears much of the cost* ***inflicted*** *by the unhealthy urban conditions in the big cities.*

Howard's ideas, although utopian, were also highly practical and were adopted around the world in the **ensuing** decades. His garden cities were intended to be planned, **self-contained** communities surrounded by parks, containing **proportionate** and separate areas of residences, industry, and agriculture. Inspired by the Utopian novel *Looking Backward* and Henry George's work *Progress and Poverty*, Howard published his book *Garden Cities of To-morrow* in 1898, commonly regarded as the most important book in the history of urban planning. His idealized garden city would house 32, 000 people on a site of 6, 000 acres (2, 428 hm^2), planned on a **concentric** pattern with open spaces, public parks and six **radial boulevards**, 120 ft (37 m) wide, extending from the centre. The garden city would be self-sufficient and when it reached full population, another garden city would be developed nearby. Howard **envisaged** a cluster of several garden cities as satellites of a central city of 50, 000 people, linked by road and rail.

He founded First Garden City, Ltd. in 1899 to create the first garden city at Letchworth, Hertfordshire. Donors to the project collected interest on their investment if the garden city generated profits through rents or, as Fishman calls the process, "**philanthropic** land **speculation**". Howard tried to include working class cooperative organizations, which included over two million members, but could not win their financial support. In 1904, Raymond Unwin' a noted architect and town planner, along with his partner Richard Barry Parker, won the competition run by the First Garden City, Limited to plan Letchworth, an area 34 miles outside London. Unwin and Parker planned the town in the centre of the Letchworth estate with Howard's large agricultural greenbelt surrounding the town, and they shared Howard's notion that the working class deserved better and more affordable housing. However, the architects **ignored** Howard's **symmetric** design, instead replacing it with a more **organic** design.

Welwyn Garden City, also in Hertfordshire was also built on Howard's principles. His successor as chairman of the Garden City Association was Sir Frederic Osborn, who extended the movement to regional planning.

The principles of the garden city were soon applied to the planning of city suburbs. The first such project was the Hampstead Garden Suburb founded by Henrietta Barnett and planned by Parker and Unwin. The scheme's utopian ideals were that it should be open to all classes of people with free access to woods and gardens and that the housing should be of low density with wide, tree-lined roads.

In North America, the Garden City movement was also popular, and **evolved** into the "Neighborhood Unit" form of development. In the early 1900s, as cars were introduced to city streets for the first time, residents became increasingly concerned with the number of **pedestrians** being injured by car traffic. The response, seen first in Radburn, New Jersey, was the Neighborhood Unit-style development, which **oriented** houses toward a common public path instead of the street. The neighborhood is distinctively organized around a school, with the intention of providing children a safe way to walk to school.

6.2 Urban planning profession

Urban planning became professionalized at this period, with input from **utopian visionaries** as well as from the practical minded infrastructure engineers and local councillors combining to produce new design **templates** for political consideration. The Town and Country Planning Association was founded in 1899 and the first academic course on urban planning was offered by the University of Liverpool in 1909.

The first official consideration of these new trends was **embodied** in the Housing and Town Planning Act of 1909 that **compelled** local authorities to introduce **coherent** systems of town planning across the country using the new principles of the garden city, and to ensure that all housing construction **conformed** to specific building standards.

Following this Act, surveyors, civil engineers, architects, lawyers and others began working together within local government in the UK to draw up schemes for the development of land and the idea of town planning as a new and distinctive area of expertise began to be formed. In 1910, Thomas Adams was appointed as the first Town Planning **Inspector** at the Local Government Board, and began meeting with practitioners. The Town Planning Institute was established in 1914 with a **mandate** to advance the study of town-planning and civic design. The first university course in America was established at Harvard University in 1924.

The Tudor Walters Committee that **recommended** the building of housing estates after World War One **incorporated** the ideas of Howard's **disciple** Raymond Unwin, who demonstrated that homes could be built rapidly and economically whilst maintaining satisfactory standards for gardens, family privacy and **internal** spaces. Unwin **diverged** from Howard by proposing that the new developments should be **peripheral** satellites rather than **fully-fledged** garden cities.

6.3 Modernism

In the 1920s, the ideas of modernism began to surface in urban planning. The influential modernist architect Le Corbusier presented his scheme for a Contemporary City for three million inhabitants (Ville Contemporaine) in 1922. The centrepiece of this plan was the group of sixty-story **cruciform** skyscrapers, steel-framed office buildings **encased** in huge curtain walls of glass. These skyscrapers were set within large, **rectangular**, park-like green spaces. At the centre was a huge transportation **hub** that on different levels included **depots** for buses and trains, as well as highway **intersections**, and at the top, an airport. Le Corbusier had the fanciful notion that commercial airliners would land between the huge skyscrapers. He **segregated pedestrian circulation** paths from the roadways and glorified the **automobile** as a means of transportation. As one moved out from the central skyscrapers, smaller low-story, **zig-zag** apartment blocks (set far back from the street amid green space) housed the inhabitants. Le Corbusier hoped that politically minded industrialists in France would lead the way with their efficient Taylorist and Fordist strategies adopted from American industrial models to re-organize society.

In 1925, he exhibited his "Plan Voisin", in which he proposed to ***bulldoze*** *most of central Paris north of the Seine and replace it with his sixty-story cruciform towers from the Contemporary City, placed within an* ***orthogonal*** *street* ***grid*** *and park-like green space.* In the 1930s, Le Corbusier **expanded** and **reformulated** his ideas on urbanism, eventually publishing them in La Ville radieuse (The Radiant City) in 1935. Perhaps the most significant difference between the Contemporary City and the Radiant City is that the latter **abandoned** the class-based **stratification** of the former; housing was now assigned according to family size, not economic position. Le Corbusier's theories were **sporadically** adopted by the builders of public housing in Europe and the United States.

Many of his **disciples** became notable in their own right, including painter-architect Nadir Afonso, who absorbed Le Corbusier's ideas into his own aesthetics theory. *Lúcio Costa's city plan of Brasília and the industrial city of Zlín planned by František Lydie Gahura in the Czech Republic are notable plans based on his ideas, while the architect himself produced the plan for Chandigarh in India.* Le Corbusier's thinking also had been **profoundly** effected by the philosophy of Futurism and Constructivism in Russia at the turn of the 20th century.

Another important theorist was Sir Patrick Geddes who understood the importance of taking the regional environment into account and the relationship between social issues and town planning, and **foresaw** the emergence of huge urban **conurbations**. In 1927, he was commissioned to plan the city of Tel Aviv, then in the British mandate for Palestine. It consisted of about 40 blocks, sized around 150 metres squared. The block contained an inner small public garden, **disposed** into a **windmill configuration** of inner access roads, making it awkward for car traffic. The big blocks form a gently **undulating** street pattern, north–south commercial, east–west arranged to catch the sea breeze. This was a simple and efficient manner to modernize the historical fixed grid patterns. A series of shaded boulevards **short** cuts the system, with some public squares, accessing the sea front. The plan of the new town became a success.

Urban planning in communist countries has often modeled itself on Western modernism, using the authority of the state to implement efficient urban designs produced in administrative centers. (In Russia this process was **nominally** decentralized after the end of the USSR, but Moscow remains the source of much of the country's urban planning expertise.) Germany under national socialism also undertook **grandiose** schemes for urban redesign.

6.4 New Towns

Ebenezer Howard's urban planning concepts were only adopted on a large scale after World War II. The damage brought on by the war **provoked** significant public interest in what post–war Britain would be like, which was encouraged by the government, who facilitated talk about a "Better Britain" to **boost morale**. Post–war rebuilding **initiatives** saw new plans drafted for London, which, for the first time, addressed the issue of **decentralization**. Firstly, the County of London Plan 1943 recognized that **displacement** of population and employment was necessary if the city was to be rebuilt at a desirable density. Moreover, the Greater London Plan of 1944 went further by suggesting that over one million people would need to be **displaced** into a mixture of satellite suburbs, existing rural towns, and new towns.

The New Towns Act 1946 resulted in many New Towns being constructed in Britain over the following decades.

New towns were built in the United States from the 1960s—examples include Reston, Virginia; Columbia, Maryland; Jonathan, Minnesota and Riverside Plaza. This construction effort was combined with extensive federal government grants for **slum** clearance, improved and increased housing and road construction and comprehensive urban renewal projects. Other European countries such as France, Germany, Italy and Sweden also had some successes with new towns, especially as part of post–war reconstruction efforts.

7 Contemporary

Urban planning has grown in **prominence** with rising urbanization.

7.1 Reaction against modernism

By the late 1960s and early 1970s, many planners felt that modernism's clean lines and lack of human scale **sapped vitality** from the community, blaming them for high crime rates and social problems.

Modernist planning fell into decline in the 1970s when the construction of cheap, **uniform** tower blocks ended in most countries, such as Britain and France. Since then many have been demolished and replaced by other housing types. Rather than attempting to **eliminate** all disorder, planning now **concentrates** on individualism and **diversity** in society and the economy; this is the post–modernist era.

Minimally planned cities still exist. Houston is a large city (with a metropolitan population of 5.5 million) in a developed country without a comprehensive **zoning ordinance**. Houston does, however, restrict development densities and **mandate** parking, even though specific land uses are not regulated. Also, private–sector developers in Houston use subdivision **covenants** and **deed** restrictions to effect land–use restrictions **resembling** zoning laws. *Houston* voters have rejected comprehensive zoning ordinances three times since 1948.

7.2 New Urbanism

Various current movements in urban design seek to create **sustainable** urban environments with long–lasting structures, buildings and a great **liveability** for its inhabitants. The most clearly defined form of **walkable** urbanism is known as the Charter of New Urbanism. It is an approach for successfully reducing environmental impacts by altering the built environment to create and **preserve** smart cities that support sustainable transport. Residents in **compact** urban neighborhoods drive fewer miles and have significantly lower environmental impacts across a range of measures compared with those living in **sprawling** suburbs. The concept of **Circular** flow land use management has also been introduced in Europe to **promote** sustainable land use patterns that strive for compact cities and a reduction of greenfield land taken by urban sprawl.

In sustainable construction, the recent movement of New Classical Architecture promotes a sustainable approach towards urban construction that **appreciates** and develops smart growth, walkability, architectural tradition, and classical design. This is in contrast to modernist and short–lived globally uniform architecture, as well as **opposing solitary** housing estates and suburban sprawl. Both trends started in the 1980s.

Critics of New Urbanism have argued that its environmental aspect is too focused on transport and **excessive** individual **mobility**. The real problem with the **unsustainable**

nature of modern cities is not just about cars and too much driving–it is about the entire urban **metabolism** of the city. They have also argued that land–use planning can do little to achieve **sustainability** without regulating the design and associated technology of the actual development within a zoned area. Distances and density are relatively unimportant; it is the total metabolism of the development that determines the environmental impact. Also, the emphasis needs to shift from sustainability to **resilience**, and the spatial scope from the city to the whole urban region. A further **criticism** is that the New Urbanist project of compacting urban form is a difficult and slow process. In the new global situation, with the horizontal, low–density growth **irreversibly** dominant, and climate change already happening, it would be wiser to focus efforts on the resilience of whole city–regions, **retrofitting** the existing sprawl for sustainability and self–sufficiency, and investing heavily in green infrastructure.

7.3 Sustainable development and sustainability

Sustainable development has emerged in recent decades as guiding themes for urban planning. This term was defined and **advocated** in 1987 report *Our Common Future*, published by the World Commission on Environment and Development.

Some planners argue that modern lifestyles use too many natural resources, polluting or destroying **ecosystems**, increasing social **inequality**, creating urban heat islands, and causing climate change. Many urban planners, therefore, advocate sustainable cities.

However, sustainable development is a recent, **controversial** concept. Wheeler, in his 2004 book, defines sustainable urban development as "development that improves the long–term social and **ecological** health of cities and towns." He sketches a sustainable city's features: compact, efficient land use; less automobile use, yet better access; efficient resource use; less pollution and waste; the restoration of natural systems; good housing and living environments; a healthy social **ecology**; a sustainable economy; community **participation** and **involvement**; and **preservation** of local culture and wisdom. *Urban planners are now **promoting** a sustainable city model, which consists of cities that are designed with consideration of environmental impacts, such as **minimizing** the uses of energy, water, and the outputs of waste and pollution.*

Because of political and governance structures in most **jurisdictions**, sustainable planning measures must be widely supported before they can affect institutions and regions. Actual **implementation** is often a complex **compromise**.

Nature in cities (Often an **integral** part of sustainable cities) is the **incorporation** of nature within a city.

Car free: sustainability in city planning can include large pedestrian zones or be a totally Car free.

7.4 Collaborative planning in the United States

Collaborative planning arose in the US in response to the **inadequacy** of traditional public participation techniques to provide real opportunities for the public to make decisions affecting their communities. *Collaborative planning is a method designed to* ***empower stakeholders*** *by elevating them to the level of decision-makers through direct engagement and dialogue between stakeholders and public agencies, to* ***solicit*** *ideas, active involvement, and participation in the community planning process.* Active public involvement can help planners achieve better outcomes by making them aware of the public's needs and preferences and by using local knowledge to inform projects. When properly administered, **collaboration** can result in more meaningful participation and better, more creative outcomes to **persistent** problems than can traditional participation methods. It enables planners to make decisions that reflect community needs and values, it **fosters** faith in the wisdom and utility of the resulting project, and the community is given a personal **stake** in its success.

Experiences in Portland and Seattle have demonstrated that successful collaborative planning depends on a number of **interrelated** factors: the process must be truly **inclusive**, with all stakeholders and affected groups invited to the table; the community must have final decision-making authority; full government **commitment** (of both financial and intellectual resources) must be **manifest**; **participants** should be given clear objectives by planning staff, who facilitate the process by providing guidance, **consultancy**, expert opinions, and research; and facilitators should be trained in **conflict** resolution and community organization.

Notes

[1] 本文源自维基百科（From Wikipedia，the free encyclopedia，https://www.wikipedia.org/）。

[2] 本文主要相关文献：李德华，2001；栾峰，2004；仇保兴，2003；沈玉麟，1989；孙施文，2007；吴志强，1999，2000；吴志强，等，2010；张京祥，2005；张庭伟，1999；周国艳，等，2010；Geddes，1915；Hall，2000；Howard，1898；Innes，1995；Jacobus，1961；Le Corbusier，1924，1933；Sager，1994；Taylor，1998；Unwin，1909；Venturi，1966；Wheeler，et al，2004。关于新城市主义，参阅：Katz，1994；Duany，et al，2000；Calthorpe，et al，2001；Dutton，2000。

[3] 课文中下划线部分为有关城市规划理论的重要论述，需强化阅读，每篇课文习题部分均要求将下划线句子或段落译成中文。

[4] 课文中下划线斜体字部分为复杂长句，其解析详见下文“**Complex Sentences**”。

New Words

abandon v. 放弃，放纵〖前缀 1. a–; 词根 18. band〗

advocate v. 拥护，支持，鼓吹；n. 支持者，拥护者〖前缀 3. ac–; 词根 232. voc〗

appreciate v. 欣赏，领会，感激

automobile n. 汽车〖前缀 9. auto–; 词根 130. mob〗

boost v. 促进，提高，增加

boulevard n. 大马路，林荫大道

bulldoze v. 推倒，铲平

champion v. 捍卫，声援

circular adj. 环形的，循环的，迂回的〖前缀 15. circu–〗

circulation n. 流通，循环〖前缀 15. circu–〗

coherent adj. 一致的，连贯的〖前缀 16. co–; 词根 90. here〗

collaboration n. 合作，协作〖前缀 16. col–; 词根 103. labor〗

collaborative adj. 合作的，协作的〖前缀 16. col–; 词根 103. labor〗

commitment n. 承诺

compact adj. 紧凑的，简洁的

compel v. 强迫，迫使〖前缀 16. com–; 词根 157. pel〗

compromise v. 妥协；危害

concentrate v. 专心于，集中〖前缀 16. con–; 词根 32. centr〗

concentric adj. 同一中心的〖前缀 16. con–; 词根 32. centr〗

configuration n. 布局，构造，配置〖前缀 16. con–; 词根 67. figure〗

conflict n. 冲突，战斗〖前缀 16. con–; 词根 73. flict〗

conform v. 符合，遵照〖前缀 16. con–; 词根 76. form〗

consultancy n. 顾问工作，咨询公司

contemporary adj. 当代的，同时代的〖前缀 16. con–; 词根 207. tempor〗

controversial adj. 有争议的，引起争议的〖前缀 17. contro–; 词根 226. vers〗

conurbation n. 有卫星城的大都市

covenant n. 协议，协定

critic n. 批评家，评论员〖词根 48. crit〗

criticism n. 批评，批判〖词根 48. crit〗

cruciform adj. “十” 字形的〖词根 76. form〗

decentralization n. 分散〖前缀 18. de–; 词根 32. centr〗

decentralize v. 分散，疏散〖前缀 18. de–; 词根 32. centr〗

deed n. 行为，行动；契约，证书

demarcate v. 定界，区分

depot n. 火车站，汽车站，航空站

dictate v. 命令，指示；控制，支配〖词根 55. dict〗

disadvantaged adj. 贫穷的，社会地位低下的〖前缀 22. dis–〗

disciple n. 信徒，门徒

displace v. 移动，移走；替代，取代〖前缀 22. dis–〗

displacement n. 取代，替代〖前缀 22. dis–〗

dispose v. 处理，处置，安排〖前缀 22. dis–; 词根 170. pos〗

diverge v. 分开，叉开〖前缀 19. di–; 词根 225. verg〗

diversity n. 多样化，多样性〖前缀 19. di–; 词根 226. vers〗

ecological adj. 生态的，生态学的〖词根 57. eco〗

ecology n. 生态学〖词根 57. eco〗

ecosystem n. 生态系统〖词根 57. eco〗

eliminate v. 排除，消除

embody v. 表现，体现〖前缀 23. em–〗

empower v. 授权，准许〖前缀 23. em–〗

encase v. 包装，围绕〖前缀 24. en–〗

ensuing adj. 接着发生的，接踵而至的

envisage v. 想象，设想〖前缀 24. en–; 词根 229. vis〗

eponymous adj. 齐名的〖词根 145. onym〗

evil n. 邪恶，罪恶，坏事

evolve v. 发展，进化〖词根 234. volv〗

excessive adj. 过度的，过分的〖前缀 26. ex–; 词根 30. cess〗

expand v. 扩张，扩展

foresaw v. 预知，预见〖前缀 28. fore–〗

formulate v. 使公式化；系统地阐述或提出〖词根 76. form〗

foster v. 培养，抚育；促进

fully–fledged adj. 羽毛丰满的

grandiose adj. 宏伟的，浮夸的〖词根 83. grand〗

grid n. 格子，方格

hub n. 中心，焦点

ignore v. 忽视，不顾〖词根 82. gnor〗

implementation n. 实施，执行

inadequacy n. 不充分；不适当〖前缀 34. in–〗

incentive n. 刺激，诱因；动机

inclusive adj. 包括的，包容广阔的〖前缀 34. in–; 词根 40. clus〗

incorporate v. 包含，吸收，合并〖前缀 34. in–; 词根 42. corpor〗

incorporation n. 合并，编入〖前缀 34. in–; 词根 42. corpor〗

inequality n. 不平等，不均等〖前缀 34. in–; 词根 58. equ〗

inflict v. 使承受，遭受〖前缀 34. in–; 词根 73. flict〗

initiate v. 发起，创始；接纳或介绍某人加入某团体等

initiative n. 主动，首创精神

inspector n. 检察员，检阅官〖前缀 34. in–; 词根 198. spec〗

integral adj. 构成整体所必需的；完整的

internal adj. 内部的〖前缀 34. in–〗

interrelated adj. 相互关联的〖前缀 35. inter–〗

intersection n. 交叉，相交，交叉点〖前缀 35. inter–; 词根 182. sect〗

intervention n. 干涉，干预〖前缀 35. inter–; 词根 222. vent〗

involvement n. 牵连，参与，加入〖前缀 34. in–; 词根 234. volv〗

irreversibly adv. 不可更改地，不可反转地〖前缀 36. ir–; 前缀 61. re–; 词根 226. vers〗

jurisdiction n. 司法权，管辖权

laissez–faire adj. 自由放任的；n. 自由放任主义

liberalism n. 自由主义〖词根 108. liber〗

liveability n.（住宅、环境的）适于居住性

mandate n. 命令，指令；v. 命令，指令〖词根 120. mand〗

manifest adj. 明白的，明显的；v. 显示，表明

metabolism n. 新陈代谢〖前缀 39. meta–〗

minimally adv. 最低限度地

minimize v. 把……减至最低数量

mitigate v. 缓和，减轻

mobility n. 流动性，移动性〖词根 130. mob〗

morale n. 士气，精神面貌

nominally adv. 名义上，表面地〖词根 137. nomin〗

oppose v. 反对，抗争〖前缀 op–, 反；词根 170. pos〗

ordinance n. 条例，法令

organic adj. 有机的，自然发展的

orient v. 标定方向，使……向东方；adj. 东方的，新生的；n. 东方〖词根

148. ori 〗

originate v. 发源，发起；创始，发明〖词根 148. ori〗

orthogonal adj. 直角的，直交的，矩形的〖词根 149. ortho〗

paradigm n. 范例，样式

participant n. 参加者，参与者

participation n. 参加，参与

pedestrian n. 步行者，行人；adj. 徒步的，平淡无奇的〖词根 156. ped〗

peripheral adj. 外围的，次要的〖前缀 54. peri-〗

persistent adj. 坚持不懈的，持久的〖前缀 53. per-; 词根 192. sist〗

philanthropic adj. 博爱的，慈善的〖词根 161. phil; 词根 12. anthrop〗

philanthropist n. 慈善家〖词根 161. phil; 词根 12. anthrop〗

preservation n. 保存，保留，保护〖词根 188. serv〗

preserve v. 保护，保持，保存〖词根 188. serv〗

profoundly adv. 深刻地，深深地

promote v. 促进，推进，提升〖前缀 58. pro-; 词根 132. mot〗

proportionate adj. 成比例的，相称的

provoke v. 激起，挑起；激怒，使愤怒〖前缀 58. pro-; 词根 232. vok〗

radial adj. 辐射状的，放射式的

rationale n. 基本原理，基础理论

reaction n. 反应；反应能力；抗拒；反动〖词根 2. act〗

recommend v. 推荐，劝告

rectangular adj. 矩形的，成直角的〖词根 177. rect〗

reformulate v. 再次阐述或提出〖前缀 61. re-; 词根 76. form〗

resemble v. 与……相像，类似于〖词根 184. semble〗

resilience n. 弹性，弹力，快速恢复的能力

retrofit v. 翻新，改型

sap n. 精力，元气；v. 使衰竭，使伤元气

segregate v. 分开，分离〖前缀 63. se-; 词根 86. greg〗

self-contained adj. 独立的，自给自足的〖前缀 64. self-〗

short adv. 简短地；横贯地

slum n. 贫民窟，贫民区

solicit v. 恳求，征求

solitary adj. 独自的，独立的〖词根 194. sol〗

speculation n. 推断；思考；投机活动

sporadically adv. 偶发地，零星地

sprawl v. 蔓延；n. 蔓延，杂乱无章拓展的地区
stake n. 股份
stakeholder n. 股东
stratification n. 分层，层化
sustainability n. 持续性，永续性
sustainable adj. 可持续的
symmetric adj. 对称的，匀称的〖前缀 70. sym-; 词根 126. metr〗
template n. 样板，模板
tremendous adj. 极大的，巨大的
undulate v. 起伏，波动
uniform adj. 一样的，规格一致的〖前缀 76. uni-; 词根 76. form〗
unsustainable adj. 不能持续的，不能支持的
utopian adj. 乌托邦的，空想的，不切实际的
visionary n. 空想家，预言家〖词根 229. vis〗
vitality n. 活力，生气，生命力
walkable adj. 适于步行的，可以走去的
windmill n. 风车
zig-zag n. "之" 字形；adj. "之" 字形的
zoning n. 分区，分带

Complex Sentences

[1] **The laissez-faire style of government management of the economy**, in fashion for most of the Victorian era, **was starting to give way to a New Liberalism** that championed intervention on the part of the poor and disadvantaged.

黑体字部分为句子的主干（主语、谓语和宾语）。in fashion for most of the Victorian era 为解释性插入语，that 引导的从句为定语从句，修饰前面的 New Liberalism。

[2] **He was also influenced by the work of economist Alfred Marshall** ***who*** *argued in 1884* ***that*** *industry needed a supply of labor that could in theory be supplied anywhere,* ***and that*** *companies have an incentive to improve workers living standards as the company bears much of the cost inflicted by the unhealthy urban conditions in the big cities.*

黑体字部分为句子的主干。整个斜体字部分为 who 引导的定语从句，修饰前面的 Alfred Marshall，其包括两个 that 引导的宾语从句，第一个 that 引导的宾语从句又可鉴别出 that 引

导的定语从句（下划线部分），修饰前面的 a supply of labor，第二个 that 引导的宾语从句又包括一个 as 引导的状语从句（下划线部分）。本句结构复杂，形成从句套从句再套从句（三重从句）的局面。

[3] In 1925, **he exhibited his "Plan Voisin"**, *in which he proposed to bulldoze most of central Paris north of the Seine and replace it with his sixty-story cruciform towers from the Contemporary City, placed within an orthogonal street grid and park-like green space.*

黑体字部分为句子的主干。斜体字部分为 which 引导的定语从句，修饰前面的 Plan Voisin，其中 placed 引导的过去分词短语作定语，修饰前面的 his sixty-story cruciform towers。

[4] **Lúcio Costa's city plan of Brasília and the industrial city of Zlín planned by František Lydie Gahura in the Czech Republic are notable plans** based on his ideas, while the architect himself produced the plan for Chandigarh in India.

本句结构较为简单,前半部分为主句,后半部分为 while 引导的状语从句。主句的主语较为复杂，注意区分人名、地名。

[5] **Urban planners are now promoting a sustainable city model,** *which consists of cities that are designed with consideration of environmental impacts, such as minimizing the uses of energy, water, and the outputs of waste and pollution.*

黑体字部分为句子的主干。斜体字部分为 which 引导的定语从句，修饰前面的 a sustainable city model，其又包含一个 that 引导的定语从句，修饰前面的 cities。

[6] **Collaborative planning is a method** ***designed to empower stakeholders*** *by elevating them to the level of decision-makers through direct engagement and dialogue between stakeholders and public agencies,* ***to solicit ideas, active involvement, and participation*** *in the community planning process.*

正体黑体字部分为句子的主干。斜体字部分为 desogned 引导的过去分词短语，作定语，修饰前面的 method。过去分词短语很长，结构较为复杂。

Exercises

[1] 将课文中下划线句子或段落译成中文。

[2] 读熟或背诵上述复杂长句。

[3] 解析下列单词［解析单词是指根据单词中包含的词根、前缀推导出单词的含义，例如，apathy（n. 漠然，冷淡），解析式为：a-（无）+ path（感情）+ y →无感情→冷漠，冷淡］。解析的单词选自上文"**New Words**"。

advocate v. 拥护，支持，鼓吹；n. 拥护者	inequality n. 不平等，不均等
automobile n. 汽车	inspector n. 检察员，检阅官
collaboration n. 合作，协作	irreversibly adv. 不可更改地，不可挽回地
compel v. 强迫，迫使	oppose v. 反对，抗争
concentric adj. 同一中心的	peripheral adj. 外围的，次要的
contemporary adj. 当代的，同时代的	persistent adj. 坚持不懈的，持久的
cruciform adj. "十"字形的	philanthropic adj. 博爱的，慈善的
decentralize v. 分散，疏散	promote v. 促进，推进，提升
eponymous adj. 齐名的	provoke v. 激起，挑起；激怒，使愤怒
excessive adj. 过度的，过分的	rectangular adj. 矩形的，成直角的
grandiose adj. 宏伟的，浮夸的	solitary adj. 独自的，独立的

Unit 4

Pioneer Thinkers in Urban Planning (1)

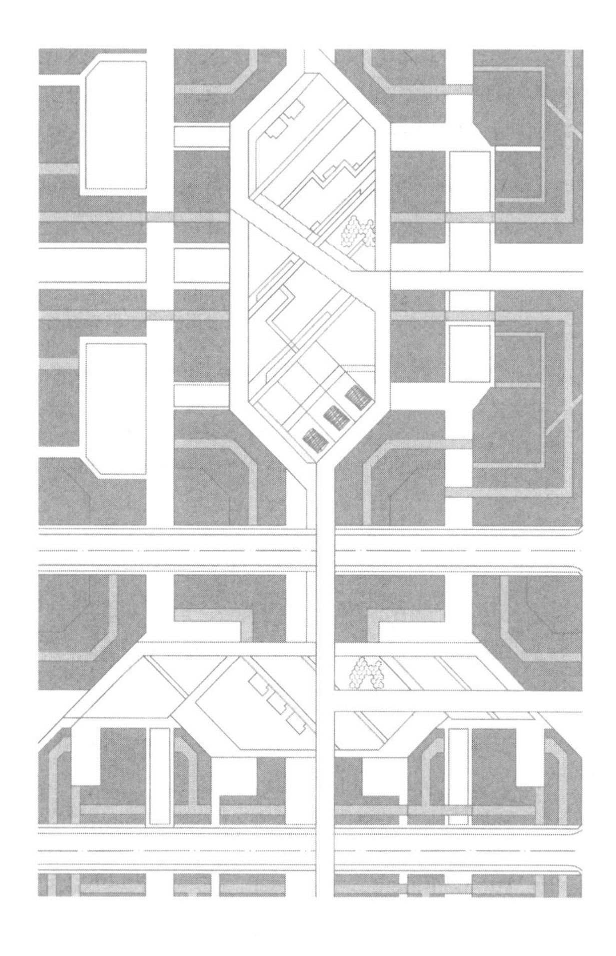

The whole of Chapter 2 has **concentrated** on the **evolution** of what can be called, broadly, the urban problem in Britain from the Industrial Revolution of the late eighteenth century to the outbreak of the Second World War. We have looked at the facts of urban development and at the attempts—often **faltering** and not very **effective** ones—on the part of central and local administration to deal with some of the resulting problems. This was the world of practical men **grappling** with practical matters. But no less important, during this time, were the writings and the influence of thinkers about the urban problem. Often their writings and their lectures reached only a tiny minority of **sympathetic** people. To practical men of the time, much of what they asserted would seem **utopian**, even **cranky**. Yet in sum, and in **retrospect**, the influence of all of them has been **literally incalculable**; furthermore, it still continues.

This delay in the recognition and acceptance of their ideas is very important. Some of these ideas were more or less fully developed by the end of the nineteenth century, and a large part were known to the interested public by the end of the First World War. Yet with the exception of some small-scale experiments up to 1939, nearly all the influence on practical policy and design has come since 1945. One obvious peril in this is that no matter how topical and how appropriate these thinkers were in analyzing the problems of their own age, their remedies might be at least partially **outdated** by the time they came to be taken seriously. We shall need to judge for ourselves how serious this has been.

It is useful to divide the thinkers into two groups: the Anglo-American group and the continental European group. The basis of the distinction here is more than one of convenience. Basically, the background of the two groups of thinkers has been quite different. *We have already seen in Chapter 2 that in England and Wales (Scotland in this respect has been rather more like the European continent), cities began to spread out after about 1860: first the middle class and then (especially with the growth of public housing after the First World War) the working class began to move out of the* ***congested*** *inner rings of cities into single-family homes with individual gardens, built at densities of 10 or 12 houses to the acre (25–30 / hectare).* Exactly the same process occurred, from about the same time, in most American cities, though in some cases the process was delayed by the great wave of arrivals of national groups (such as Italians, Greeks, Russians, Poles, and Jews from Russia and Poland) between 1880 and 1910; they crowded together in **ethnic ghettos** in the inner areas of cities like New York, Boston and Chicago, and took some time to join the general outward movement. Nevertheless, by the 1920s and 1930s there was a rapid growth of single-family housing around all American cities, served by public transport and then, increasingly, by the private car. This was a tradition which, by and large, writers and thinkers in both Britain and the United States accepted as the starting point.

On the Continent it was quite otherwise. As cities grew rapidly under the impact of industrialization and movement from the countryside, generally several decades after the **equivalent** process had taken place in Britain (i.e. from about 1840 to 1900), they failed to spread out to anything like the same extent. As public transport services developed, generally in the form of horse and then electric tram systems, some of the middle class, and virtually all the working class, continued to live at **extraordinarily** high densities virtually within walking distance of their work. The typical Continental city consisted then, and still consists today, of high apartment blocks—four, five or six storeys high—built continuously along the streets, and thus **enclosing** a big **internal** space within the street block. In middle-class areas this might be a pleasant communal green space; in other areas it was **invariably** built over in the **desperate** attempt to crowd in as many people as possible. The result by 1900 was the creation of large **slum** areas in most big European cities, but of a form quite different from the English slums. In England even poor people lived in small—generally two-storey—houses of their own, either rented or bought. In continental Europe they lived in small apartments, and the densities—in terms of both persons per room and dwellings or persons per net residential acre—were much higher than in typical English slum areas. (Scotland, **curiously** developed in the European way: Glasgow, for instance, is a city of tenements, not houses, and standards of crowding have always been much worse there than in big English cities.) Naturally, when continental Europeans began to think about urban planning, they tended to accept as a starting point this apparent preference for high-density apartment-living within the city.

1 The Anglo-American tradition (1)

1.1 Ebenezer Howard

The first, and without doubt the most influential, of all the thinkers in the Anglo American group is Ebenezer Howard (1850-1928). His book *Garden Cities of Tomorrow* (first published in 1898 under the title *To-morrow*, and republished under its better-known title in 1902) is one of the most important books in the history of urban planning. Reprinted several times and still readily available as a **paperback**, it remains astonishingly topical and relevant to many modern urban problems. From it stems the whole of the so-called garden city (or in modern **parlance**, new town) movement which has been so influential in British urban planning theory and practice.

To understand its significance it is necessary to look at its historical background. Howard was not a professional planner—his career, if he can be said to have had one, was as a **shorthand** writer in the law courts—but a private individual who liked to **speculate**, write and organize. As a young man he travelled, spending a number of years in the United States during its period of rapid urban growth before returning to England to write his book. At that time several pioneer industrialists with **philanthropic** leanings had already started new communities in association with large new factories which they had built in open countryside. (Their motives, perhaps, were not entirely philanthropic: they built their factories cheaply on rural land; it was necessary to house the labor force outside the city in consequence, and they got a modest return in rents for their investment.) The earliest of these experiments, Robert Owen's **celebrated** experimental settlement at New Lanark in Scotland (c. 1800-1810) and Titus Salt's town built round his textile mill at Saltaire near Bradford (1853-1863), actually date from the early years of the Industrial Revolution. But the best known and the most important date from the late nineteenth century, when the growing scale of industry was tending to **throw up** a few very powerful industrialists who saw the advantages of **decentralizing** their plants far from the existing urban congestion. Bournville, outside Birmingham (1879-1895), built by the chocolate manufacturer George Cadbury, and Port Sunlight on the Mersey near Birkenhead (1888), built by the chemical **magnate** William Hesketh Lever, are the best-known examples in Britain. In Germany the engineering and **armaments** firm of Krupp built a number of such settlements outside its works at Essen in the Ruhr district, of which the best preserved, Margarethenhöhe (1906), closely **resembles**

Bournville and Port Sunlight. Similarly, in the United States the railroad engineer George Mortimer Pullman (who invented Pullman cars) built a model town named after himself, outside Chicago, from 1880 onwards.

These towns all contain the germ of the idea which Howard was to **propagate**: in all of them industry was decentralized **deliberately** from the city, or at least from its inner sections, and a new town was built around the decentralized plant, thus combining working and living in a healthy environment. They are, in a sense, the first garden cities, and many of them are still functional and highly pleasant towns today. But Howard generalized the idea from a simple company town, the work of one industrialist, into a general planned movement of people and industry away from the crowded nineteenth century city. Here he drew on previous writings: on Edward Gibbon Wakefield, who had **advocated** the planned movement of population even before 1850, and James Silk Buckingham, who had developed the idea of a model city. But perhaps the strongest intellectual influence on Howard's thinking was that of the great Victorian economist Alfred Marshall; he, if anyone, invented the idea of the new town as an answer to the problems of the city, and he gave it an economic justification which only later came to be fully understood. Marshall argued, as early as 1884, that much industry was even then **footloose**, and would locate anywhere if labor was available; he also recognized that the community would eventually have to pay the social costs of poor health and poor housing, and that these were higher in large cities (as they then existed) than they would be in new model communities.

Howard, however, developed the idea, generalized it and above all turned it into an **eminently** practical call for action. And of all **visionary** writers on planning, Howard is the least **utopian**, in the sense of **impractical**; his book is packed with detail, especially financial detail, of how the new garden cities were to be built. But first of all Howard had to provide a justification of the case for new towns (or garden cities) that could be readily understood by practical men without much knowledge of economics. He did so in the famous diagram of the Three Magnets, which in fact is an extremely compressed and brilliant statement of planning objectives. (It is an interesting exercise to try to write out the diagram in suitably **jargon-ridden**, abstract modern language as a statement of objectives; to say the same thing less clearly takes many pages, whereas Howard got it all in one simple diagram.) Basically, Howard was saying here that both existing cities and the existing countryside had an **indissoluble** mixture of advantages and disadvantages. The advantages of the city were the opportunities it offered in the form of **accessibility** to jobs and to urban services of all kinds; the disadvantages could all be summed up in the poor resulting natural environment. **Conversely**, the countryside offered an excellent environment but virtually no opportunities of any sort.

It is important here to remember the date of Howard's book. In the 1890s material conditions in British cities were better than they had been in the 1840s. Average incomes for many workers were significantly higher; medical standards had improved; and the new housing by-laws were beginning to have effect. Nevertheless, by modern standards they were still **appalling**. The 1891 census showed that at least 11 per cent of the population, over 3 million people, were living at densities of over 2 persons per room; and this was certainly an **underestimate**. Even in the 1880s the Registrar General's records showed that the expectation of life in a city like Manchester was only 29 years at birth on average—only 5 years more than forty years previously. In the late 1880s and early 1890s the shipowner Charles Booth conducted the first modern social investigation, based on strict statistical recording. Aided by the young Beatrice Webb, he produced a study which is still a classic: it showed that on a strict and minimal standard **no less than** one-quarter of the population of inner east London was living below the poverty line. But on the other hand there was equal **distress** in the countryside: these were the years of deep **agrarian depression**, brought about by the mass importation of cheap foreign meat and wheat, against which the British farmer was given no protection. The population map of Britain in the 1890s shows losses almost everywhere except for the limited areas of the cities and the industrial districts. Though the towns were beginning to spawn suburbs, there was virtually none of the twentieth-century phenomenon whereby urban workers could afford to live in the countryside; that had to wait for the motor car. And when Howard's book was published, it was precisely two years since Parliament had removed the requirement that a car must be **preceded** along the highway by a man with a red flag. He could, perhaps, hardly be expected to have foreseen the consequences of **liberating** the car.

Against this background Howard argued that a new type of settlement—Town Country, or Garden City—could uniquely combine all the advantages of the town by way of accessibility, and all the advantages of the country by way of environment, without any of the disadvantages of either. This could be achieved by planned decentralization of workers and their places of employment, thus **transferring** the advantages or urban **agglomeration en bloc** to the new settlement. (In modern economic **jargon**, this would be called **"internalizing** the **externalities"**.) The new town so created would be deliberately outside normal **commuter** range of the old city. It would be fairly small—Howard suggested 30,000 people—and it would be surrounded by a large green belt, easily accessible to everyone. Howard advised that when the town was established, 6,000 acres (2,400 hm^2) should be purchased: of this, no less than 5,000 acres (2,000 hm^2) would be left as green belt, the town itself occupying the remainder.

Two important points about Howard's idea especially need stressing, because they have been so widely **misunderstood**. *The first is that* ***contrary*** *to the usual impression Howard was advocating quite a high residential density for his new towns: about 15 houses per acre (37 / hectare), which in terms of prevailing family size at the time meant about 80–90 people per acre (200–220 / hectare). (Today it would mean 40–50.)* The second is that he did not advocate small, **isolated** new towns. His notion was that when any town reached a certain size, it should stop growing and the excess should be **accommodated** in another town close by. Thus the settlement would grow by cellular addition into a complex **multicentred** agglomeration of towns, set against a green background of open country. (And even this was to be fairly densely populated by space consuming urban activities like public institutions: Howard allowed for one person to every 4 acres (2 hm^2) there.) Howard called this **polycentric** settlement the "Social City". The diagram in the first edition of his book showed it as having a population of 250,000–the original target population of the modern "giant" new town of Milton Keynes in England–but Howard himself stressed that the Social City could grow without limit. This point has never been well understood, because the second edition of Howard's book, and all **subsequent** editions, have omitted the diagram; it is reproduced here in Figure 3.2.

Howard, as we have noticed, was very specific about how his new communities could be built. Private enterprise could do it, he stressed, if money could be borrowed for the purpose: land could be bought cheaply in the open countryside for the project, and the subsequent increase in land values would allow the new town company to repay the money in time and even make a profit to be **ploughed** back into further improvement, or into the creation of further units of the Social City. In fact, Howard was actually **instrumental** in getting two garden cities started: Letchworth in northern Hertfordshire (1903) and Welwyn Garden City a few miles to the south (1920). Both were built very much on the lines he advocated, with wide green belts around. But both suffered financial troubles, and the vision of private enterprise new towns on a large scale was never realized. Furthermore, despite insistent and effective **propaganda** from the Town and Country Planning Association, which he founded, governments after the First World War failed to respond to the call for public new towns.

1.2 Raymond Unwin and Barry Parker

Between 1900 and 1940 many of Howard's ideas were developed by his faithful followers. Among the most **prolific** and brilliant of the writers was Sir Frederic Osborn (1885–1978), who lived to see over a score of new towns built in England after the Second

World War. In terms of physical realization the opportunities were clearly more limited. The two architects who designed the first garden city, Letchworth, Raymond Unwin (1863–1940) and his young assistant Barry Parker (1867–1947), later went on to build Hampstead Garden Suburb at Golders Green in north-west London (1905–1909). As its name indicates, this was not a garden city but a dormitory suburb owing its existence to the new underground line opened in the year 1907; and it was **condemned** by many garden city supporters on that ground. But it was an interesting experiment in the creation of a socially mixed community, with every type of house from the big mansion to the small **cottage**; and in its creation of a range of houses which are all skillfully designed, all varied yet all quietly **compatible**, it is one of the triumphs of twentieth century British design.

Later, Parker went on to a more **ambitious** enterprise: the design of a new community for 100,000 people to be built by the City of Manchester at Wythenshawe, south of the city (1930). Wythenshawe in fact deserves to be called the third garden city (or new town) actually started in Britain before the Second World War. It has all the essential features of the design of Letchworth or Welwyn: the surrounding green belt, the mixture of industrial and residential areas, and the emphasis on single-family housing of good design. (The family **resemblance** between Hampstead Garden Suburb and Wythenshawe is more than **coincidental**.) It did, however, **compromise** on the principle of **self-containment**: because most of its inhabitants came from the city, where they held jobs, **subsidized** public transport was provided for them to **commute** back. But the intention—never completely realized in practice—was to provide a wide range of jobs in the community itself.

Together, Unwin and Parker developed some important modifications of the original Ebenezer Howard idea. In a very influential pamphlet published in 1912, *Nothing Gained by Overcrowding!* Unwin argued that housing should be developed at lower densities than were then common. The need for public open space, he pointed out, was related to the numbers of people, so that the saving in land from higher urban densities was largely **illusory**. He recommended a net density in new residential areas of about 12 houses to the acre (30 / hectare)—or, in terms of the average family size of the time, about 50–60 people to the acre (124–150 / hectare). This standard was accepted in the important official Tudor Walters Report of 1918, as we have seen, and became usual in most public housing schemes of the 1920s and 1930s. Wythenshawe, like many other major schemes by city housing departments, was built at about this density.

Both Unwin and Parker consistently argued for the Howard principle of generous green belts around the new communities. In Unwin's graphic term, used in the regional plan he produced for the London area in the late 1920s, they would be cities against a background of open space—not cities surrounded by green belts, in the conventional use of the term. But Parker developed the idea still further. Visiting the United States in the 1920s, he was impressed by the early experiments in building parkways, i.e. scenic roads running through landscaped open country. Parker argued that the "background of open space" between cities should be occupied by these parkways, giving easy **interconnection** between them; this in fact was an adaptation to the motor age of Howard's original idea of an inter–urban railway. Parker actually managed to half–build one parkway in the middle of Wythenshawe, and it was later completed as the M56 north Cheshire motorway, though not as he would have intended it.

Lastly, at Wythenshawe, Parker employed yet another notion he had picked up in the United States, which was in fact a logical development of Howard's own ideas: the idea of dividing the town into clearly ***articulated*** *neighborhood units.* The ground plan of Wythenshawe as actually completed shows the influence of this idea. To see its **origins**, we now need to follow across the Atlantic the Anglo–American tradition of thought.

Notes

[1] 本文选自：Hall P. Urban and Regional Planning[M]. 4th edition. London and New York: Routledge, 2002. 彼得·霍尔. 城市和区域规划（原著第四版）[M]. 邹德慈，李浩，陈熳莎，译. 北京：中国建筑工业出版社，2008。

[2] 本文主要相关文献：Beevers, 1987; Cherry, 1981; Creese, 1992; Fishman, 1977; Friedmann, et al, 1979; Geddes, 1915; Hall, 2002; Howard, 1898; Jackson, 1985; Le Corbusier, 1924, 1933; Meller, 1990; Reiner, 1963; Stalley, 1972; Sutcliffe, 1981; Tetlow, 1968; Unwin, 1909; Wright, 1935。

[3] 课文中下划线部分为有关城市规划理论的重要论述，需强化阅读，每篇课文习题部分均要求将下划线句子或段落译成中文。

[4] 课文中下划线斜体字部分为复杂长句，其解析详见下文"**Complex Sentences**"。

New Words

accessibility n. 可达性〖前缀 3. ac–; 词根 30. cess〗

accommodate v. 容纳，向……提供；使适应，使和谐一致

advocate v. 拥护，支持，鼓吹；n. 支持者，拥护者〖前缀 3. ad–; 词根 232. voc〗

agglomeration n. 成团，结块

agrarian adj. 农村的，农业的，土地的

ambitious adj. 有雄心的，有野心的〖前缀 5. ambi–〗

appalling adj. 骇人听闻的，令人震惊的

armament n. 军备，武器，装备

articulate v. 清楚说话；接合

celebrated adj. 著名的，有名的

coincidental adj. 巧合的，同时发生的〖前缀 16. co–; 词根 23. cid〗

commute v. 交换；坐公交车上下班〖前缀 16. com–; 词根 135. mut〗

commuter n. 乘公交车辆上下班者，经常乘车往返者〖前缀 16. com–; 词根 135. mut〗

compatible adj. 相容的，兼容的〖前缀 16. com–; 词根 153. pati〗

compromise v. 妥协；危害

concentrate v. 集中，聚集，专心于〖前缀 16. con–; 词根 32. centr〗

condemn v. 谴责；判刑

congest v. 充满，拥挤

contrary adj. 相反的，反对的，对立的〖前缀 17. contra–〗

conversely adv. 相反地〖前缀 16. con–; 226. vers〗

cottage n. 小屋，村舍

cranky adj. 古怪的

curiously adv. 好奇地；奇怪地

decentralize v. 分散，疏散〖前缀 18. de–; 词根 32. centr〗

deliberately adv. 深思熟虑地，故意地〖前缀 18. de–; 词根 108. liber〗

depression n. 沮丧，萎靡不振〖前缀 18. de–; 词根 172. press〗

desperate adj. 绝望的；孤注一掷的；急切的，极度渴望的

distress n. 悲痛，不幸，贫困

effective adj. 有效的

eminently adv. 突出地，显著地

en bloc adj. & adv. 全部的（地）

enclose v. 把……围起来〖前缀 24. en–〗

equivalent adj. 相等的，相当的；n. 对等物〖词根 58. equ〗

ethnic adj. 种族的，部落的

evolution n. 演变，进化，发展〖词根 234. volu〗

externality n. 外在性，外形，外部事物〖词根 26. ex–〗

extraordinarily adv. 特别地，不平常地〖词根 27. extra–〗

faltering adj. 犹豫的，支吾的

footloose adj. 未婚的，独身的，自由自在的

ghetto n. 少数民族的集中住宅区，贫民区

grapple v. 抓住

illusory adj. 虚幻的，虚无缥缈的

impractical adj. 不切实际的〖前缀 33. im–〗

incalculable adj. 数不清的，不可估量的〖前缀 34. in–〗

indissoluble adj. 不能分解的〖前缀 34. in–; 词根 196. solu〗

instrumental adj. 有帮助的，起作用的

interconnection n. 相互连接〖前缀 35. inter–〗

internal adj. 内部的，体内的，国内的〖前缀 34. in–〗

internalize v. 使内在化，使藏在心地〖前缀 34. in–〗

invariable adj. 恒定的，不变的，始终如一的〖前缀 34. in–; 词根 221. vari〗

isolate v. 隔离，孤立〖词根 194. sol〗

jargon n. 行话，行业术语

liberate v. 解放，释放〖108. liber〗

literally adv. 逐字地，照字面地；确实地，真正地〖词根 110. liter〗

magnate n. 富豪，权贵，巨头，大资本家〖词根 118. magn〗

misunderstood adj. 被误解的〖前缀 41. mis–〗

multicentred adj. 多中心的〖前缀 43. multi–; 词根 32. centr〗

no less than 不少于，不下于，多达

origin n. 起源，根源；出身〖词根 148. ori〗

outdate v. 过时〖前缀 47. out–〗

paperback n. 平装本（书）

parlance n. 说法，用语

philanthropic adj. 博爱的，慈善的〖词根 161. phil; 词根 12. anthrop〗

plough v. 用犁耕田，开路，投资

polycentric adj. 多中心的〖前缀 55. poly–; 词根 32. centr〗

precede v. 在……之前发生或出现，先于〖前缀 57. pre–; 词根 30. cede〗

prolific adj. 多产的，多结果的〖词根 109. life〗

propaganda n. 宣传，宣传活动

propagate v. 繁衍，增值

resemblance n. 相似，形似〖词根 184. semble〗

resemble v. 与……相像，类似于〖词根 184. semble〗

retrospect n. & v. 回顾，回想〖前缀 62. retro–; 词根 198. spec〗

ridden adj.（通常构成复合词）充满的

self-contained adj. 独立的，自给自足的〖前缀 64. self–〗

shorthand n. 速记（法）

slum n. 贫民窟

speculate v. 思索，推测，投机
subsequent adj. 随后的，跟随的〖前缀 65. sub–; 词根 187. sequ〗
subsidize v. 补助，给……补贴
sympathetic adj. 同情的，赞同的〖前缀 70. sym–; 词根 153. path〗
throw up 产生（人才）
transfer v. 使转移，使调动；n. 转移，调动〖前缀 71. trans–〗
underestimate n. 低估，看轻〖前缀 75. under–〗
utopian adj. 乌托邦的，空想的，不切实际的
visionary adj. 预言性的，有远见的；空想的；n. 预言家；空想家〖词根 229. vis〗

Complex Sentences

[1] **We have already seen** in Chapter 2 *that in England and Wales (Scotland in this respect has been rather more like the European continent),* ***cities began to spread out*** *after about 1860*: ***first the middle class and then*** *(especially with the growth of public housing after the First World War)* ***the working class began to move*** *out of the congested inner rings of cities into single–family homes with individual gardens, built at densities of 10 or 12 houses to the acre (25–30 / hectare).*

正体黑体字部分为句子的主干（主语和谓语）。斜体字部分为 that 引导的宾语从句及其解释性分句（冒号前、后）；冒号前斜体黑体字部分为宾语从句的主干，冒号后斜体黑体字部分为解释性分句的主干；冒号前 in 和 after 引导的介词短语在宾语从句中作状语，冒号后 out of...into... 结构介词短语在解释性分句中作状语，with 引导的介词短语以及 built 引导的过去分词短语在解释性分句中作定语，修饰前面的 single–family homes；括号内容为解释性插入语。

[2] Against this background **Howard argued** *that* ***a new type of settlement***—*Town Country, or Garden City*—***could uniquely combine all the advantages of the town*** *by way of accessibility,* ***and all the advantages of the country*** *by way of environment, without any of the disadvantages of either.*

正体黑体字部分为句子的主语和谓语。斜体字部分为 that 引导的宾语从句，斜体黑体字部分为宾语从句的主干，破折号之间为解释性插入语，against，by 以及 without 引导的介词短语均作状语。

[3] **The first is** *that contrary to the usual impression* ***Howard was advocating quite a high residential density for his new towns***: ***about 15 houses per acre (37 / hectare),*** *which in terms of prevailing family size at the time meant about 80–90 people per acre (200–220 / hectare). (Today it would mean 40–50.)*

正体黑体字部分为句子的主语和谓语。斜体字部分为 that 引导的表语从句，斜体黑体字部分为表语从句的主干，contrary 引导的形容词短语作状语，which 引导的定语从句修饰前面的 about 15 houses per acre (37 / hectare)，括号内为解释性插入语。

[4] Lastly, at Wythenshawe, **Parker employed yet another notion** he had picked up in the United States, which was in fact a logical development of Howard's own ideas: the idea of dividing the town into clearly articulated neighborhood units

黑体字部分为句子的主干（主语、谓语和宾语），he had picked up in the United States 为省略关系代词 that 或 which 的定语从句，修饰前面的 notion; which 引导的从句亦为定语从句，亦修饰前面的 notion；冒号后内容为对 notion 的进一步解释。

Exercises

[1] 将课文中下划线句子或段落译成中文。

[2] 读熟或背诵上述复杂长句。

[3] 解析下列单词 [解析单词是指根据单词中包含的词根、前缀推导出单词的含义，例如，apathy（n. 漠然，冷淡），解析式为：a-（无）+ path（感情）+ y →无感情→冷漠，冷淡]。解析的单词选自上文"**New Words**"。

accessibility n. 可达性	isolate v. 隔离，孤立
coincidental adj. 巧合的，同时发生的	magnate n. 富豪，权贵，巨头，大资本家
commute v. 交换；坐公交车上下班	philanthropic adj. 博爱的，慈善的
contrary adj. 相反的，反对的，对立的	polycentric adj. 多中心的
conversely adv. 相反地	precede v. 在……之前发生或出现，先于
decentralize v. 分散，疏散	prolific adj. 多产的，多结果的
deliberately adv. 深思熟虑地，故意地	retrospect n. & v. 回顾，回想
depression n. 沮丧，萎靡不振	subsequent adj. 随后的，跟随的
equivalent adj. 相等的，相当的；n. 对等物	sympathetic adj. 同情的，赞同的
indissoluble adj. 不能分解的	underestimate n. 低估，看轻

Unit 5

Pioneer Thinkers in Urban Planning (2)

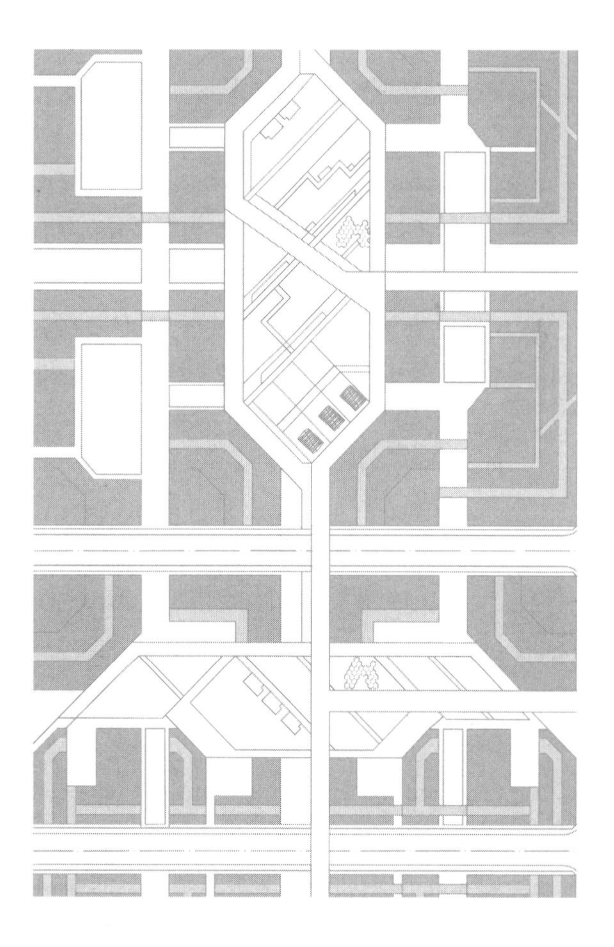

2 The Anglo-American tradition（2）

2.1 Clarence Perry, Clarence Stein and H. Alker Tripp

In Howard's original theoretical diagram of his Garden City, published in 1898, he divided the town up into "wards" of about 5,000 people, each of which would contain local shops, schools and other services. *This, in* ***embryo****, is the origin of the neighborhood unit idea, which in essence is merely* ***pragmatic****: certain services, which are provided every day for groups of the population who cannot or do not wish to travel very far (housewives and young children), should be made available at an* ***accessible*** *central place for a fairly small local community, within walking distance of all homes in that community.* Depending on the residential density, the idea of convenient walking distance will **dictate** a limit of a few thousand people for each of these units. It makes psychological sense to give such a unit a clear identity for the people who live in it, by arranging the houses and streets so that they focus on the central services and providing some obvious boundary to the outside.

In the United States, however, the idea was taken much further during the preparation of the *New York Regional Plan* in the 1920s. (This great multi-volume plan, prepared wholly by a voluntary organization, is one of the milestones of twentieth century planning.) One contributor to this plan, Clarence Perry (1872-1944), there developed the idea of the neighborhood unit, not merely as a pragmatic device, but as a **deliberate** piece of social engineering which would help people achieve a sense of identity with the community and with the place. (He based it on a model garden suburb, Forest Hills Gardens in New York

City, which he had helped to plan in 1912.) For this there was no empirical justification, and not much has emerged since; though some important work done for the Royal Commission on Local Government in England, and published in 1967, did suggest that most people's primary sense of identification was to a very small local area. But just in terms of physical planning Perry's work did give **firmness** to the neighborhood idea. He suggested that it should consist of the **catchment** area of a primary school, **extending** about half or three-quarters of a mile in any direction and containing about 1,000 families–or about 5,000 people, in terms of average family size then. It would be bounded by main traffic roads, which children should not be expected to cross.

The essential idea of the neighborhood unit, as developed by Perry–though not some of his details, such as putting shops at the corners of the units at the **junctions** of traffic roads–was **enthusiastically** taken up by British planners in new towns and in some cities after the Second World War; its influence is everywhere to be seen. Since the early 1960s, however, it has come in for increasing criticism. An influential paper published in 1963 by Christopher Alexander, a young English **emigre** to the United States, called *A City Is Not a Tree*, suggested that sociologically the whole idea was false: different people had varied needs for local services, and the principle of choice was **paramount**. In his view cities that had grown naturally demonstrated a more complex settlement structure, with overlapping fields for shops and schools; planners should aim to reproduce this variety and freedom of choice. The master plan for the new town of Milton Keynes, published in 1969, was one of the first to reflect these ideas.

Meanwhile, a close associate of Perry–Clarence Stein (1882–1975), an architect-planner working in the New York region—had taken the neighborhood concept further. Stein was one of the first physical planners, apart from Parker in England and Le Corbusier in France, to face fully the **implications** of the age of mass ownership of the private car. He **grasped** the principle that in local residential areas the need above all was to **segregate** the **pedestrian** routes used for local journeys—especially by housewives and children—from the routes used by car traffic. In a new town development at Radburn, northern New Jersey (1933), which was started but never completed, he applied these ideas by developing a **separate** system of pedestrian ways, reached from the back doors of the houses, which pass through communal open-space areas between the houses, and thence cross under the vehicle streets. The vehicle streets, in turn, are designed according to a **hierarchical** principle, with main primary routes giving **access** to local distributors and then in turn to local access roads designed on the **cul-de-sac** (dead end) principle, serving small groups of houses. The Radburn Layout, as it came to be known, was applied by Stein in one or two other developments in the United States in the 1930s, but was adopted in Britain only after the Second World War; in fact, most of the examples date from the late 1950s or later. Few of them have the charm and ease of the design in the surviving section of Radburn itself.

The **transfer** of Stein's ideas back across the Atlantic to Britain, there to be combined with Perry's neighborhood idea, came via a curious route. In 1942 an imaginative Assistant Commissioner of Police (Traffic) at London's Scotland Yard, called H. Alker Tripp (1883–1954), published a slim book called *Town Planning and Traffic*. Though there is no direct evidence that Tripp knew of Perry's or Stein's work, it seems possible that he had read of it. The most novel suggestion in the book was the idea that after the war British cities should be **reconstructed** on the basis of **precincts**. *Instead of main city streets which served mixed functions and which had many points of access to local streets, thus giving rise to congestion and accidents, Tripp argued for a* ***hierarchy*** *of roads in which main* ***arterial*** *or sub–arterial roads were sharply* ***segregated*** *from the local streets, with only occasional access, and also were free of direct* ***frontage*** *development.* These high–capacity, free–flow highways would define large blocks of the city, each of which would have its own shops and local services. Tripp illustrated the idea graphically in his book by applying it to an **outworn** (and heavily bombed) section of London's East End.

Tripp's book came at an opportune time. For at that point Patrick Abercrombie—the most notable professional planner in Britain of that age—was working with the chief architect of the London County Council, J.H. Forshaw, and his brilliant assistant Wesley Dougill, on a postwar reconstruction plan for London. In an important section of the plan (which was published in 1943) Abercrombie and Forshaw called for the widespread application of the precinctual principle to London, and illustrated its application to two critical areas where traffic threatened the urban life and **fabric**: the zone around Westminster Abbey, and the university quarter of western Bloomsbury. **Ironically**, in neither area were the ideas ever applied; indeed, at the end of the 1950s a new one–way traffic scheme actually routed through traffic through the heart of Abercrombie's Bloomsbury precinct. But elsewhere—most notably in the postwar reconstruction of the centre of Coventry, one of Britain's most heavily bombed cities—the idea was employed to good effect.

2.2 Patrick Geddes and Patrick Abercrombie

Abercrombie's most notable contributions to Anglo–American planning theory and practice, however, were made in extending city planning to a wider scale—the scale which embraced the city and the whole region around it in a single planning exercise. To understand this tradition and the way Abercrombie fits into it, it is necessary to go back some way and consider how the Ebenezer Howard tradition developed in another, slightly different direction.

From 1883 to 1919 a **visionary** Scots biologist, Patrick Geddes (1854–1932), taught at the University of Dundee, but did much of his most important work at his famous outlook tower in Edinburgh. Geddes's extraordinary mind soon took him away from conventional biology into the area we should now recognize as human ecology: the relationship between humans and their environment. In turn he was led to a systematic study of the forces that were shaping growth and change in modern cities, which **culminated** in his masterpiece *Cities in Evolution* (published in 1915, but mostly written about 1910). To understand the nature of Geddes's achievement, as with Howard's, it is necessary to place the book in the context of its time. Human geography, which had developed so finely in France during the first decade of the twentieth century in the hands of such practitioners as Vidal de la Blache and Albert Demangeon, was an almost unknown study in Britain; only one man, H.J. Mackinder, kept the subject alive, at the University of Oxford. But Geddes was fully **acquainted** with this tradition of study, and with the associated work of the French sociologist P.G.F. le Play. Both stressed the **intimate** and subtle relationships which existed between human settlement and the land, through the nature of the local economy; in le Play's famous **triad**, the relationship Place–Work–Folk was the fundamental study of people living in and on their land.

Geddes's contribution to planning was to base it firmly on the study of reality: the close analysis of settlement patterns and local economic environment. This led him to go right outside the conventional limits of the town, and to stress the natural region—a favorite unit of analysis of the French geographers—as the basic framework for planning. Today, when so many students are trained in the basic principles of human geography at school, all this seems very obvious and familiar. But, published at a time when planning for most practitioners was the study of civic design at a quite local level—a sort of applied architecture—it was quite revolutionary. Howard had already anticipated the change of scale; his analysis of the problem, and its solution, was a regional one. Geddes's contribution was to put the **flesh** of reality on the bare bones of the regional idea: at last, human geography was to provide the basis of planning. From this came Geddes's working method, which became part of the standard sequence of planning: survey of the region, its characteristics and trends, followed by analysis of the survey, followed only then by the actual plan. Geddes, more than anyone, gave planning a logical structure.

But his contribution did not end there. His analysis of cities in evolution led him to what was then a **novel** conclusion. Suburban **decentralization**, we saw in Chapter 2, was already by then causing cities to spread more widely. But in addition, certain basic locational factors—

the pull of coalfields in the early nineteenth century, the natural **nodality conferred** on certain regions by the way railways, roads and canals followed natural routes, the economies of scale and **agglomeration** in industry—had already caused a marked **concentration** of urban development in certain regions, such as the West Midlands, Lancashire, central Scotland in Britain, or the Ruhr coalfield in Germany. Geddes demonstrated that in these regions suburban growth was causing a tendency for the towns to **coalesce** into giant urban agglomerations or conurbations—the first time this word was used in the English language.

The conclusion Geddes drew was a logical one: if this was happening and would continue to happen, under the pressure of economic and social forces, town planning must be **subsumed** under town and country planning, or planning of whole urban regions encompassing a number of towns and their surrounding spheres of influence. Howard and his supporters had already drawn the same conclusion; and between the two world wars, aided powerfully by the persuasive writing of Geddes's American follower Lewis Mumford—whose 1938 text *The Culture of Cities* became almost the **bible** of the regional planning movement—the idea gained a great deal of **credence** among thinking planners and administrators. Unwin himself was **commissioned** to prepare an advisory plan for London and its region, though funds ran out in the **depression** of 1931–1933 before it could properly be completed. Already, here, Unwin was applying the ideas of Howard to a planned scheme for large-scale decentralization of people and jobs from London to satellite towns in the surrounding Home Counties.

In this plan can be seen the germ of *the Greater London Plan of 1944*, which Patrick Abercrombie (1879–1957) prepared at the direct request of the British government. (It is significant that in this case, as elsewhere, extraordinary arrangements had to be made to prepare even an advisory plan for a whole urban region; the existing machinery of local government was quite **inadequate** in scale for the purpose.) *But Abercrombie's great achievement was to* ***weld*** *this complex of ideas, from Howard through Geddes to Unwin, and turn them into a graphic blueprint for the future development of a great region—a region centered on the metropolis but extending for 30 miles (50 km) around it in every direction, and encompassing over 10 million people.* The broad aim of the plan was essentially Howard's: it was the planned decentralization of hundreds of thousands of people from an overcrowded giant city and their re-establishment in a great series of new planned communities, which from the beginning would be self-contained towns for living and working. The method was essentially Geddes's survey of the area as it was, including the historical trends which could be observed, followed by systematic analysis of the problem, followed by production of the plan. But the great sweep of the study, its characteristic assurance, and its quality of almost cartoon-like clarity were essentially Abercrombie's own. However, *the Greater London Plan* essentially belongs to the story of the development of British planning at the time of the Second World War, which I treat in Chapter 4. We shall save a full discussion of it until then.

2.3 Frank Lloyd Wright

For the last important figure in the Anglo-American tradition we have to return across the Atlantic. Frank Lloyd Wright (1869–1959) does not fit readily into the line of development outlined in the **previous** pages, or into any line at all. It is fitting to put him last in this series, because his ideas about urban planning are so fundamentally at **variance** with those of the Continental school. Above all, they stand at the opposite extreme from those of Le Corbusier—the only other master of modern architecture whose ideas on planning are significant.

As with Corbusier, Wright's best-known monuments are his individual buildings, several of which are milestones of the modern movement; his ideas for planning on a wider scale never got further than paper. But unlike those of Corbusier, Wright's ideas were never taken up **enthusiastically** by a large following, either in Europe or in his native United States. They have, however, continued to exercise an important hold on a few influential thinkers in American planning practice during the 1950s and 1960s, especially in California. This is just, because in the same way that Corbusier's ideas are **quintessentially** European, so Wright's are typically American.

Wright based his thinking on a social **premise**: that it was desirable to preserve the sort of independent rural life of the **homesteaders** he knew in Wisconsin around the 1890s. To this he added the realization, based on the early spread of the motor car among the farmers of North America, that mass car use would allow cities to spread widely into the countryside. With the car and with cheap electric current everywhere, Wright argued, the old need for activities to concentrate in cities had ended: **dispersion**, not only of homes but also of jobs, would be the future pattern. He proposed to accept this and to encourage it by developing a completely dispersed—though planned—low-density urban spread, which he called Broadacre City. Here, each home would be surrounded by an acre of land, enough to grow crops on; the homes would be connected by superhighways, giving easy and fast travel by car in any direction. Along these highways he proposed a planned roadside civilization, in which the petrol ("gas") station would grow naturally into the **emporium** for a whole area; thus he anticipated the out-of-town shopping centre some twenty years before it actually arrived in North America. In fact, Wright's description of Broadacre City proved to be an **uncannily** accurate picture of the typical settlement form of North America after the Second World War—except that today the big 1/2-acre or 1-acre (0.2–0.4 hm^2) lots grow very little food to support their families. The form developed without the underlying social basis that Wright so **devoutly** hoped for.

Notes

[1] 本文选自:Hall P. Urban and Regional Planning[M]. 4th edition. London and New York: Routledge, 2002. 彼得·霍尔.城市和区域规划（原著第四版）[M]. 邹德慈，李浩，陈熳莎，译.北京：中国建筑工业出版社，2008。

[2] 本文主要相关文献：Beevers, 1987; Cherry, 1981; Creese, 1992; Fishman, 1977; Friedmann, et al, 1979; Geddes, 1915; Hall, 2002; Howard, 1898; Jackson, 1985; Le Corbusier, 1924, 1933; Meller, 1990; Reiner, 1963; Stalley, 1972; Sutcliffe, 1981; Tetlow, 1968; Unwin, 1909; Wright, 1935。

[3] 课文中下划线部分为有关城市规划理论的重要论述，需强化阅读，每篇课文习题部分均要求将下划线句子或段落译成中文。

[4] 课文中下划线斜体字部分为复杂长句，其解析详见下文“**Complex Sentences**”。

New Words

access n. 接近，进入，通路；v. 接近，使用〖前缀 3. ac–; 词根 30. cess〗
accessible adj. 易达到的，易受影响的〖前缀 3. ac–; 词根 30. cess〗
acquaint v. 使熟悉，使认识
agglomeration 成团，结块
arterial adj. 动脉的
bible n. 圣经
catchment n. 集水，汇水；集水处，汇水处
coalesce v. 联合，合并
commission v. 委任，授予
concentration n. 集中；浓度〖前缀 16. con–; 词根 32. centr〗
confer v. 授予，颁予
credence n. 相信〖46. cred〗
cul-de-sac n. 死胡同
culminate v. 达到极点
decentralization n. 分散〖前缀 18. de–; 词根 32. centr〗
deliberate adj. 故意的，深思熟虑的〖前缀 18. de–; 词根 108. liber〗

depression n. 沮丧，萎靡不振〖前缀 18. de-; 词根 172. press〗

devoutly adv. 虔诚地

dictate v. 命令，指示；控制，支配〖词根 55. dict〗

dispersion n. 分散，散开

embryo n. 胚胎

emigre n. 移居国外的人，移民；逃亡者，流亡者

emporium n. 商场，商业中心

enthusiastically adv. 热心地，热情地

extend v. 延伸，扩大，推广〖前缀 26. ex-; 词根 208. tend〗

fabric n. 织物；构造，结构

firmness n. 坚定，坚固

flesh n. 肉，肉体

frontage n. 正面，前面

grasp v. 抓住

hierarchical adj. 等级的，分层的〖词根 14. arch(y)〗

hierarchy n. 分层，层次，等级制度〖词根 14. arch(y)〗

homesteader n. 农场所有权者，自耕农

implication n. 暗示，含意；卷入，牵连〖前缀 33. im-; 词根 169. pli〗

inadequate adj. 不充足的；不适当的〖前缀 34. in-〗

intimate adj. 亲密的，亲近的

ironically adv. 嘲讽地，挖苦地，具有讽刺意味地

junction n. 连接，结合；会合点，结合点〖词根 101. junct〗

nodal adj. 节的，结的

novel adj. 新颖的，新奇的〖词根 141. nov〗

outworn adj. 用旧的，废弃的〖前缀 47. out-〗

paramount adj. 最重要的，具有最高权力的〖前缀 51. para-; 词根 133. mount〗

pedestrian n. 步行者，行人；adj. 徒步的；平淡无奇的〖词根 156. ped〗

pragmatic adj. 实际的，实用主义的

precincts n. 区，管辖区

premise n. 前提

previous adj. 先前的，以前的〖前缀 57. pre-〗

quintessentially adv. 典型地，标准地

reconstruct v. 重建，改造〖前缀 61. re-; 词根 201. struct〗

segregate v. 分离，隔离〖前缀 63. se-; 词根 86. greg〗

separate adj. 分开的，分隔的〖前缀 63. se-〗

subsume v. 归入，包括

transfer v. 使转移，使调动；n. 转移，调动〖前缀 71. trans-〗

triad n. 三个一组

uncannily adv. 惊异地，神秘地

variance n. 变化，变动〖词根 221. vari〗

visionary adj. 预言性的，有远见的；空想的；n. 预言家；空想家〖词根 229. vis〗

weld v. 焊接，使紧密结合

Complex Sentences

[1] **This,** in embryo, **is the origin of the neighborhood unit idea,** ***which in essence is merely pragmatic***: ***certain services,*** *which are provided every day for groups of the population who cannot or do not wish to travel very far (housewives and young children),* ***should be made available*** *at an accessible central place for a fairly small local community, within walking distance of all homes in that community.*

正体黑体字部分为句子的主干（主语、谓语和表语）。斜体字部分为 which 引导的定语从句及其解释性分句（以冒号为分隔）。冒号后斜体黑体字部分为解释性分句的主干，which 引导的定语从句修饰其前面的 certain services，who 引导的定语从句修饰其前面的 groups of the population, 括号内为解释性插入语。

[2] *Instead of main city streets which served mixed functions and which had many points of access to local streets, thus giving rise to congestion and accidents,* **Tripp argued for a hierarchy of roads** in which main arterial or sub-arterial roads were sharply segregated from the local streets, with only occasional access, and also were free of direct frontage development.

黑体字部分为句子的主干（主语、谓语和宾语），其后为 which 引导的结构复杂的定语从句，修饰其前面的 a hierarchy of roads。Instead of 引导的介词短语作状语，其又包括两个 which 引导的定语从句，修饰前面的 main city streets。thus giving rise to congestion and accidents 为解释性插入语。

[3] But **Abercrombie's great achievement was to weld this complex of ideas,** from Howard through Geddes to Unwin, **and turn them into a graphic blueprint for the future development of a great region**–a region centered on the metropolis but extending for 30 miles (50 km) around it in every direction, and encompassing over 10 million people.

黑体字部分为句子的主干（主语、谓语和表语，动词不定式作表语，双表语）。from Howard through Geddes to Unwin 为解释性插入语。破折号后亦为解释性插入语，只是这一插入语结构复杂，包括三个作定语的分词短语（一个过去分词短语，两个现在分词短语）。

Exercises

[1] 将课文中下划线句子或段落译成中文。

[2] 读熟或背诵上述复杂长句。

[3] 解析下列单词[解析单词是指根据单词中包含的词根、前缀推导出单词的含义，例如，apathy（n. 漠然，冷淡），解析式为：a-（无）+ path（感情）+ y →无感情→冷漠，冷淡]。解析的单词选自上文“**New Words**”。

concentration n. 集中；浓度	paramount adj. 最重要的，具有最高权力的
credence n. 相信	previous adj. 先前的，以前的
dictate v. 命令，指示；控制，支配	reconstruct v. 重建，改造
extend v. 延伸，扩大，推广	segregate v. 分离，隔离
hierarchy n. 分层，层次，等级制度	transfer v. 使转移，使调动；n. 转移，调动
implication n. 暗示，含意；卷入，牵连	variance n. 变化，变动

Unit 6

Pioneer Thinkers in Urban Planning (3)

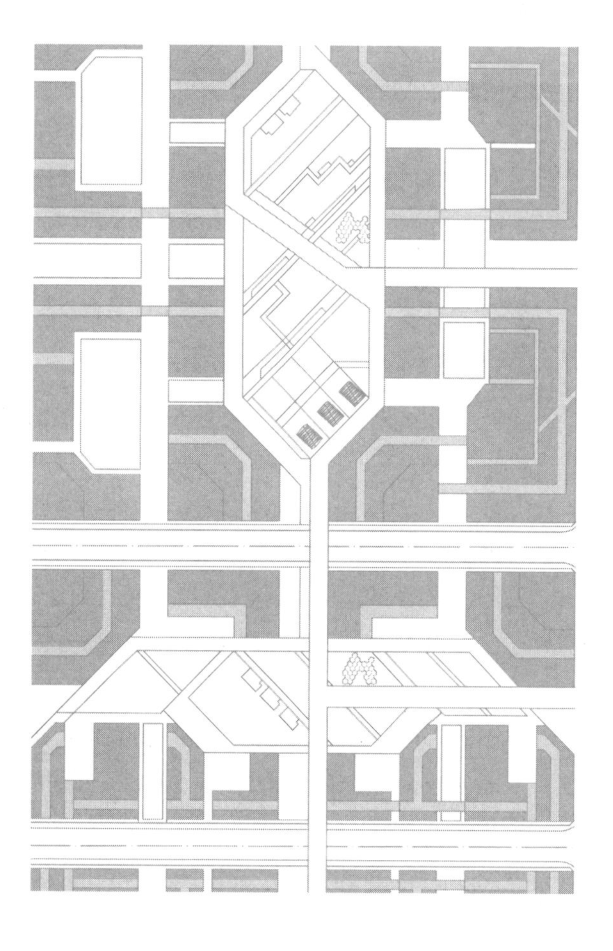

3 The European tradition

Planning as a tradition in Europe goes back to the ancient Greeks. In the nineteenth century it produced such **celebrated** designs as the reconstruction of Paris under Georges Eugène Haussmann (1809–1891), which **imposed** a new pattern of broad **boulevards** and great parks on the previous **labyrinthine** street pattern. But since our task here is to understand how new ideas **transformed** town planning into city–regional planning, we again look at the **visionaries**.

3.1 Arturo Soria y Mata

Like many other thinkers considered here, the first representative of the opposing European tradition, the Spanish engineer Arturo Soria y Mata (1844–1920), owes his place in history to the importance of one basic idea. In 1882 he proposed to develop a linear city (La Ciudad lineal), to be developed along an axis of high–speed, high–intensity transportation from an existing city. His argument was that under the influence of new forms of mass transportation, cities were tending to assume such a linear form as they grew—an argument which, as we already saw in Chapter 2, had some **justification** at that time. Soria y Mata's ideas were **ambitious** if nothing else: he proposed that his linear city might run across Europe from Cadiz in Spain to St Petersburg in Russia, a total distance of 1,800 miles (2,900 km). In fact, he succeeded only in building a few kilometers just outside Madrid; these still survive, though they are difficult to pick out on the map or on the ground, because they have been **swallowed** up in the **amorphous** growth of the modern city. In some ways the form seems **archaic** today: a main road runs straight through the linear centre of the city, carrying a **tramway**, with rather **geometrical** housing blocks on either side. And there seems no doubt from experience that such a form is difficult and costly to build; furthermore, even though **commuter** journeys may be fast, they are certainly likely to be long.

Nevertheless, the idea has always enjoyed some popularity among planners on the grounds that it has some good qualities. It does correspond to the need to exploit costly investments in new lines of rapid communication, whether these are nineteenth-century railways or twentieth-century motorways. (In both these cases, though, the settlement form that is most likely to arise is not linear, but rather a series of **blobs** round the stations or interchanges on the high-speed route.) And it does give easy **access** to nearby open countryside. Furthermore, it can respond **automatically** to the need for further growth, by simple addition at the far end; it does not need to operate through **restrictive** green belts, as Ebenezer Howard's finite garden has to. So it is not surprising that the form has often appeared in regional plans as the most obvious **alternative** to the Howard-Abercrombie tradition. Parker's parkway is one example; the well-known MARS plan for London (1943), produced by a group of architects, used it; variants of it, in the postwar period, have appeared in plans for Copenhagen (1948), Washington (1961), Paris (1965) and Stockholm (1966). Some of these plans are discussed in more detail in Chapter 7. But one point can be made here: both in Washington and in Paris it proved extremely difficult to preserve the plan in the face of private attempts to build in the spaces left between the fingers or axes of urban growth. The claim that the linear city is a natural form, therefore, does not seem justified.

3.2 Tony Garnier and Ernst May

The garden city was soon **exported** across the Channel. But **curiously**, in France its best-known expression seems to have occurred **spontaneously**, at about the same time that Howard was writing, without any **mutual interaction**. Tony Garnier (1869-1948), an architect working mainly in the city of Lyon, produced in 1898—the same year precisely as Howard's was published—a design for an industrial city (Cité industrielle) which, like Howard's garden city, was to be a **self-contained** new settlement with its own industries and housing close by. The actual site Garnier chose to illustrate his scheme was just outside Lyon, and on it Garnier placed a rather strange **elongated** town, developed on a linear grid. The site plan, then, was **unoriginal**; but the detailing, with single-family houses in their own gardens, was remarkable for the France of that time. Furthermore, and perhaps the most striking feature, Garnier's houses made full use of new techniques of concrete construction—anticipating, in some ways, the designs of Le Corbusier over twenty years later.

The Cité industrielle was never built, and though garden cities were built around Paris during the 1920s and 1930s, most of them **diverged** in practice from their English models. They contained a high proportion of apartment blocks, and were remarkable mainly for their freer use of open space in the form of squares and public parks. Yet even this, given the **appallingly congested** and unhealthy quality of Parisian working class housing at that time, represented a great achievement.

In Germany too the garden city movement took early root, and there it produced some interesting results. The most notable was in the city of Frankfurt am Main, where in the 1920s the city planner and architect Ernst May (1886–1970) developed a series of satellite towns on open land outside the built–up limits, and separated from the city proper by a green belt. These were not true garden cities on the Howard model, for most workers had to **commute** into the city—in that respect they somewhat **resembled** the Wythenshawe development in Manchester at about the same time—but they were remarkable for their detailed design treatment, in which May combined **uncompromising** use of the then new functional style of architecture with a free use of low–rise apartment blocks, all set in a park landscape. Though the original form of the satellite towns has almost been lost in the growth of the city since the Second World War, this detailed design survives to impress the **contemporary** visitor. Only Berlin, where Martin Wagner was city planner, achieved the same consistently high level of design in its satellite cities during those years.

3.3 Le Corbusier

The Swiss–born architect Charles Edouard Jeanneret (1887–1965), who early in his professional career adopted the **pseudonym** Le Corbusier, stands with Frank Lloyd Wright, Walter Gropius and Mies van der Rohe as one of the creators of the modern movement in architecture; and among the general public his fame is probably greater than that of any of the others. *Yet though his best–known achievements consist of an astonishing range of individual buildings all stamped* ***indelibly*** *with his personality, from the Villa Savoye at Passy (1929–1930) to the chapel of Notre Dame en Haut at Ronchamp near Belfort (1950–1953), his most outstanding contribution as a thinker and writer was as an urban planner on the grand scale.* Of the scores of designs which Corbusier produced for city reconstructions, or for new settlements—both in France, where he worked all his professional life, and widely across the world—few materialized. The most notable are his Unité d'Habitation (1946–1952) at Marseilles in France, and his grand project for the capital city of the Punjab at Chandigarh (1950–1957), which is being finished only long after his death.

His central ideas on planning are contained in two important books, *The City of Tomorrow* (1922) and *The Radiant City* (*La Ville radieuse,* 1933), which is available in English translation. Unfortunately, Corbusier does not translate or summarize easily. The words pour out in no particular logical order, accompanied by diagrams which often contain the real sense of what is being said; the books seem to consist of collections of papers put together on no consistent principle; the style is highly **rhetorical**, and often even **declamatory**. But in so far as it is possible to make a very summary digest, his ideas seem to reduce themselves into a small number of propositions.

The first was that the traditional city has become functionally **obsolete**, owing to its increasing size and increasing congestion at the centre. As the urban mass grew through **concentric** additions, more and more strain was placed on the communications of the innermost areas, above all the central business district, which had the greatest accessibility and where all businesses wanted to be. Corbusier's classic instance, often quoted, was Manhattan Island with its skyscrapers and its congestion.

The second was the **paradox** that the congestion could be cured by increasing the density. There was a key to this, of course: the density was to be increased at one scale of analysis, but decreased at another. Locally, there would be very high densities in the form of massive, tall structures; but around each of these a very high proportion of the available ground space—Corbusier **advocated** 95 per cent—could and should be left open. The landscape he advocated, which can be seen in countless of his writings from his Paris plan of 1922 onwards, consists, therefore, of skyscrapers separated by very large areas of **intervening** open space. Thus Corbusier is able to achieve the feat of very high overall densities—with up to 1,000 people to the net residential acre (2,500 / hectare) and more—while leaving the **bulk** of the ground unbuilt on.

The third proposition concerned the distribution of densities within the city. Traditionally, as we noted in Chapter 2, densities of residential population are higher in the centre of the city than at the edge. Since the development of mass urban transportation from the 1860s onwards, the "density gradient" has **flattened** somewhat, with lower densities at the centers and rather higher densities farther out than the rural densities which used to obtain; but it is quite noticeable nevertheless, and in continental European cities (as well as some American cities, such as New York) it is much more pronounced than in Britain. Furthermore, there is an even more pronounced gradient of employment density, with big surviving concentrations near the centre. Corbusier proposed to do away with all this by **substituting** virtually equal densities all over the city. This would reduce the pressure on the central business districts, which would in effect disappear. Flows of people would become much more even across the whole city, instead of the strong radial flows into and out of the centre which characterize cities today.

Fourth and lastly, Corbusier argued that this new urban form could **accommodate** a new and highly efficient urban transportation system, **incorporating** both railway lines and completely **segregated elevated** motorways, running above the ground level, though, of course, below the levels at which most people lived. Corbusier even claimed to have invented, in the early 1920s, the multi-level free-flow highway **interchange**, long before such structures were built in Los Angeles or elsewhere.

To yield the full promised results, and thus to be open to testing in practice, Corbusier's plans would need to have been applied on a very wide scale. His own diagrams show large areas of Paris, including historic quarters, **razed** to accommodate the new forms. This is one good reason why it proved so difficult to **execute** any of his ideas—especially in interwar France, where the pace of physical construction was extremely **sluggish**—and his notions about density have seldom been applied anywhere in the extreme he suggested. Corbusier himself became increasingly **frustrated** by his failure to get his plans implemented, and he began to call for an **autocrat** like Louis XIV or Napoleon III, who would have the boldness to execute his ideas. Nevertheless, in the planning of cities after the Second World War, Corbusier's general influence has been **incalculable**. A whole generation of architects and planners trained in the 1930s and then from 1945 onwards came to revere the writings of "Corbu"; and in practice afterwards, they tried to apply his ideas to local conditions. *In England, for instance, his influence was particularly strong in the famous London County Council (LCC) Architect's Department during the 1950s, at a time when it produced much of its best work: the celebrated Alton West estate at Roehampton in south-west London (1959), with multi-storey blocks set among areas of finely landscaped parkland, is completely Corbusian in concept.* All over Britain the remarkable change in the urban landscape during the late 1950s and the 1960s—as **slum** clearance and urban renewal produced a sudden **unprecedented** crop of skyscrapers—is a **mute tribute** to Corbusier's influence. Whether it was for good or ill, later generations will have to decide. Certainly, by the end of the 1960s there was an increasing volume of protest at the **inhumanity** of the new high blocks; and it seemed doubtful whether many more would be built. Many critics were going further, and questioning the whole philosophy of massive urban renewal which was essential to the realization of Corbusier's ideas.

Corbusier has, however, had another, more subtle influence. Though many of his ideas were **intuitive** rather than scientifically exact, he did teach planners in general the importance of scale in analysis. The notion that densities could be varied locally to produce very different results, while maintaining the overall density **unaltered**, was a very simple yet at the same time **elusive** one, which few grasped fully before he demonstrated it. **Equally** important was his **insistence** on the elementary truth that dense local concentrations of people help support a viable, frequent mass-transportation system. This realization, for instance, has been extremely important in the much-admired Stockholm suburbs built in the post-1950 period, where densities are systematically higher around the new underground railway stations than they are farther away.

But in general the basic difference between Anglo-American and continental European traditions has **persisted**. The two lines have **intermingled** more in the post1945 period than ever before, it is true: in many urban renewal areas of British cities it would really be difficult at first glance to tell whether one was in Birmingham (or Newcastle), Amsterdam, Milan or Warsaw. For many **bourgeois** homebuyers and even some planners on the Continent there has been **enthusiasm** for the English idea of single-family home living and the creation of new communities in the countryside. Nevertheless, the majority of British people still appear to prefer a single-family home with garden if given the choice, while many people in Continental countries are quite firmly **wedded** to the advantages of inner-city apartment living.

4 A verdict on the seers

It may seem difficult, on first impression, to pass a general **verdict** on a group of planners as varied as those considered in this chapter. But in the light of the distinctions made in Chapter 1 of this book, it is possible to draw some conclusions that apply to almost all of them.

The first point is that most of these planners were concerned with the production of blueprints, or statements of the future end state of the city (or the region) as they desired to see it; in most cases they were far less concerned with planning as a continuous process which had to accommodate subtle and changing forces in the outside world. Their vision seems to have been that of the planner as **omniscient** ruler, who should create new settlement forms, and perhaps also destroy the old, without interference or question. *The complexities of planning in a mixed economy where private interests will **initiate** much of the development that actually occurs, or in a **participatory democracy** where individuals and groups have their own, often **contradictory** notions of what should happen—all these are absent from the writings of most of these pioneers.*

Howard and Geddes are, perhaps, honorable exceptions to most of this criticism. Howard's idea may have seemed **utopian**, but he never avoided the practical details of how to bring it about. Geddes, even more, was **explicitly** concerned that planning should start with the world as it is, and that it should try to work with trends in the economy and society, rather than **impose** its own **arbitrary** vision of the world. It is perhaps significant that his intellectual background was different from that of many of the others. An architect, by definition, starts thinking in terms of the structures s/he would like to build; a biologist turned geographer–sociologist starts by thinking about the nature of the society and the land s/he is planning for.

This leads us to a second point about most of the pioneers. Their blueprints seldom admitted of alternatives. There was one true vision of the future world as it ought to be, and each of them saw himself as its **prophet**. This is understandable, because these men were visionaries trying to be heard in a **skeptical** and sometimes **hostile** world. But if the idea is too persuasive, there is an evident risk of **stifling orthodoxy**.

One last point will be very evident. These pioneers were very much physical planners. They saw the problems of society and of the economy in physical terms, with a physical or spatial solution in terms of a particular arrangement of bricks and **mortar**, steel and concrete on the ground. This again is understandable; they were trained to think in this way and their concerns were with physical development. Nevertheless, this attitude carries with it a real peril: that such planners, and those they teach and influence, will come to see all problems of cities and regions as capable of solution in these terms and only these terms. According to this view, problems of social **malaise** in the city will be met by building a new environment to replace the old—**whereupon** poor health, **inadequate** education, badly balanced diets, marital **discord** and juvenile **delinquency** will all go away. Similarly, problems of **circulation** and traffic congestion in the city will be dealt with by designing a radical new system as part of a new urban form—whereupon, of course, the problems will disappear. The notion that not all problems are capable of simple solution in these physical terms—or the more disturbing notion that there might be cheaper or better solutions to the problems, of a non–physical character—is not often found in the writings of the pioneers of planning thought we have been discussing. Nor, it should be noted, is it often found in the plans of many of those countless planners these men have influenced and **inspired**. The seers have made their mark as much by their limitations as by their positive qualities—striking though these latter may have been.

Notes

[1] 本文选自：Hall P. Urban and Regional Planning[M]. 4th edition. London and New York: Routledge, 2002. 彼得 · 霍尔 . 城市和区域规划（原著第四版）[M]. 邹德慈，李浩，陈熳莎，译 . 北京：中国建筑工业出版社，2008。

[2] 本文主要相关文献：Beevers, 1987; Cherry, 1981; Creese, 1992; Fishman, 1977; Friedmann, et al, 1979; Geddes, 1915; Hall, 2002; Howard, 1898; Jackson, 1985; Le Corbusier, 1924, 1933; Meller, 1990; Reiner, 1963; Stalley, 1972; Sutcliffe, 1981; Tetlow, 1968; Unwin, 1909; Wright, 1935。

[3] 课文中下划线部分为有关城市规划理论的重要论述，需强化阅读，每篇课文习题部分均要求将下划线句子或段落译成中文。

[4] 课文中下划线斜体字部分为复杂长句，其解析详见下文“**Complex Sentences**”。

New Words

access n. 接近，进入，通路；v. 接近，使用〖前缀 3. ac-; 词根 30. cess〗

accommodate v. 容纳，向……提供；使适应，使和谐一致

advocate v. 拥护，支持，鼓吹；n. 支持者，拥护者〖前缀 3. ad-; 词根 232. voc〗

alternative adj. 择一的，供选择的；n. 选择，替换物〖词根 7. alter〗

ambitious adj. 有雄心的，有野心的〖前缀 5. ambi-〗

amorphous adj. 无定形的〖前缀 1. a-; 词根 131. morph〗

appallingly adv. 令人毛骨悚然地

arbitrary adj. 主观的，武断的，专横的

archaic adj. 古代的，过时的，陈旧的〖前缀 8. arch-〗

autocrat n. 独裁统治者，独断专行的人〖前缀 9. auto-; 词根 44. crat〗

automatically adv. 自动地〖前缀 9. auto-〗

blob n. 黏糊糊的一团

boulevard n. 大马路，林荫大道

bourgeois adj. 资产阶级的，资本家的

bulk n. 大多数，大部分

celebrated adj. 有名的，著名的

circulation n. 流通，循环〖前缀 15. circu-〗

commute v. 交换；坐公交车上下班〖前缀 16. com-; 词根 135. mut〗

commuter n. 乘公交车辆上下班者，经常乘车往返者〖前缀 16. com-; 词根 135. mut〗

concentric adj. 同一中心的〖前缀 16. con-; 词根 32. centr〗

congest v. 充满，拥挤

contemporary adj. 当代的，同时代的〖前缀 16. con–; 词根 207. tempor〗

contradictory adj. 矛盾的；反驳的〖前缀 17. contra–; 词根 55. dict〗

curiously adv. 好奇地；奇怪地

declamatory adj. 慷慨激昂的，夸夸其谈的〖前缀 18. de–; 词根 38. clam〗

delinquency n. 不法行为

democracy n. 民主〖词根 54. dem; 词根 44. cracy〗

discord n. & v. 不和，不一致〖前缀 20. dis–; 词根 41. cord〗

diverge v. 分开，分叉，分歧〖前缀 19. di–; 225. 词根 verg〗

elevate v. 将某人或某物举起〖词根 106. lev〗

elongate v. 延长，加长

elusive adj. 难以捉摸的；逃避的

enthusiasm n. 热情

equally adv. 相等地，平等地，公正地〖词根 58. equ〗

execute v. 执行，履行，完成

explicitly adv. 明白地，明确地〖前缀 26. ex–; 词根 169. pli〗

export v. 出口，输出〖前缀 26. ex–〗

flatten v. 变平

frustrate v. 挫败，使受挫折

geometrical adj. 几何的，几何学的〖词根 geo, 地；词根 126. metr〗

hostile adj. 敌人的，敌对的，怀有敌意的

impose v. 强加；征税〖前缀 33. im–; 词根 170. pos〗

inadequate adj. 不充足的；不适当的〖前缀 34. in–〗

incalculable adj. 数不清的，不可估量的〖前缀 34. in–〗

incorporate v. 合并，并入〖前缀 34. in–; 词根 42. corpor〗

indelibly adv. 不能消灭地

inhumanity n. 无情，残忍〖前缀 34. in–; 词根 93. hum〗

initiate v. 开始，发起

insistence n. 坚持

inspire v. 鼓舞，激励，启迪〖前缀 34. in–; 词根 200. spir〗

interaction n. 相互作用，相互影响〖前缀 35. inter–; 词根 2. act〗

interchange v. 互换，交换；n. 互通式立体交叉，道路立体枢纽〖前缀 35. inter–〗

intermingle v. 混合〖前缀 35. inter–〗

intervene v. 干涉，介入〖前缀 35. inter–; 词根 222. ven〗

intuitive adj. 知觉的，直观的

justification n. 正当理由；辩护〖词根 102. just〗
labyrinthine adj. 迷宫似的，曲折的
malaise n. 不适，不舒服
mortar n. 砂浆
mute adj. 无声的，缄默的
mutual adj. 相互的，彼此的
obsolete adj. 废弃的，已过时的
omniscient adj. 无所不知的，博识的〖前缀 46. omni–; 词根 180. sci〗
orthodoxy n. 正统，正教〖词根 149. ortho〗
paradox n. 矛盾
participatory adj. 提供参加机会的，供人分享的
persist v. 坚持〖前缀 53. per–; 词根 192. sist〗
prophet n. 预言家
pseudonym n. 假名，笔名〖前缀 60. pseudo–; 词根 145. onym〗
raze v. 彻底摧毁，将……夷为平地
resemble v. 与……相像，类似于〖词根 184. semble〗
restrictive adj. 限制的，限定的，约束的
rhetorical adj. 修辞的；辞藻华丽的，夸张的
segregate v. 分离，隔离〖前缀 63. se–; 词根 86. greg〗
self–contained adj. 独立的，自给自足的
skeptical adj. 怀疑的
sluggish adj. 行动迟缓的，反应慢的
slum n. 贫民窟
spontaneously adv. 自然地，自发地
stifling adj. 令人窒息的，沉闷的
substitute v. & n. 代替，替换
swallow v. 吞下，咽下；吞没，淹没，使消失
tramway n. 有轨电车，有轨电车路线
transform v. 改变，改观〖前缀 71. trans–; 词根 76. form〗
tribute n. 赞辞；贡物〖词根 214. tribut〗
unaltered adj. 未改变的，未被改变的〖前缀 74. un–; 词根 7. alter〗
uncompromising adj. 不妥协的，坚定的〖前缀 74. un–〗
unoriginal adj. 非独创的，无创造性的〖前缀 74. un–; 词根 148. ori〗
unprecedented adj. 前所未有的〖前缀 74. un–; 前缀 57. pre–; 词根 30. cede〗
utopian adj. 乌托邦的，空想的，不切实际的

verdict n. 裁决，裁定〖词根 223. ver; 词根 55. dict〗

visionary adj. 预言性的，有远见的；空想的；n. 预言家；空想家〖词根 229. vis〗

wedded adj. 渴望……的，执着于……的

whereupon adv. & conj. 于是，因此

Complex Sentences

[1] *Yet though his best-known achievements consist of an astonishing range of individual buildings all stamped indelibly with his personality, from the Villa Savoye at Passy (1929–1930) to the chapel of Notre Dame en Haut at Ronchamp near Belfort (1950–1953),* **his most outstanding contribution** as a thinker and writer **was as an urban planner** on the grand scale.

黑体字部分为句子的主干。斜体字部分为句子的状语从句，其中 stamped 引导的过去分词短语作定语，修饰前面的 an astonishing range of individual buildings, from...to... 结构介词短语亦作定语，亦修饰前面的 an astonishing range of individual buildings。

[2] In England, for instance, **his influence was particularly strong** in the famous London County Council (LCC) Architect's Department during the 1950s, at a time when it produced much of its best work: ***the celebrated Alton West estate at Roehampton in south-west London (1959)****, with multi-storey blocks set among areas of finely landscaped parkland,* ***is completely Corbusian in concept.***

正体黑体字部分为句子的主干（主语、谓语和表语），其后三个介词短语（分别由 in, during 和 at 引导）均作状语，第三个介词短语 (at 引导的介词短语）包含一个 when 引导的定语从句，修饰前面的 time。冒号后斜体字部分为解释性分句，斜体黑体字部分为解释性分句的主干，with 引导的介词短语和 set 引导的过去分词短语均作定语，修饰前面的 the celebrated Alton West estate。

[3] **The complexities of planning** in a mixed economy **where** private interests will initiate much of the development **that** actually occurs, or in a participatory democracy **where** individuals and groups have their own, often contradictory notions of **what** should happen—**all these are absent from the writings of most of these pioneers.**

The complexities of planning 即 all these，为句子的主语，其被其后两个 in 引导的介词短语修饰（作定语），第一个介词短语包括两个定语从句（where 和 that 引导），第二个介词短语包括两个定语从句（where 和 what 引导），结构非常复杂。

Exercises

[1] 将课文中下划线句子或段落译成中文。

[2] 读熟或背诵上述复杂长句。

[3] 解析下列单词[解析单词是指根据单词中包含的词根、前缀推导出单词的含义，例如，apathy（n. 漠然，冷淡），解析式为：a-（无）+ path（感情）+ y →无感情→冷漠，冷淡]。解析的单词选自上文“**New Words**”。

amorphous adj. 无定形的	explicitly adv. 明白地，明确地
archaic adj. 古代的，过时的，陈旧的	incorporate v. 合并，并入
autocrat n. 独裁统治者，独断专行的人	interaction n. 相互作用，相互影响
circulation n. 流通，循环	intervene v. 干涉，介入
concentric adj. 同一中心的	omniscient adj. 无所不知的，博识的
contemporary adj. 当代的，同时代的	orthodoxy n. 正统，正教
declamatory adj. 慷慨激昂的，夸夸其谈的	persist v. 坚持
democracy n. 民主	pseudonym n. 假名，笔名
discord n. & v. 不和，不一致	transform v. 改变，改观
diverge v. 分开，分叉，分歧	unprecedented adj. 前所未有的
elevate v. 将某人或某物举起	verdict n. 裁决，裁定

Unit 7

Theories of Urban Planning

Planning theory is the body of scientific concepts, definitions, behavioral relationships, and assumptions that define the body of knowledge of urban planning. There are ten procedural theories of planning that remain the principal theories of planning procedure today: the **blueprint** planning, the **synoptic** planning, the **participatory** planning, the **incremental** planning, the mixed **scanning** model, the **transactive** planning, the **advocacy** planning, the radical planning, the **bargaining** model and the **communicative** planning.

1 Background

The modern **origins** of urban planning lie in the movement for urban reform that arose as a **reaction** against the disorder of the industrial city in the mid–19th century. Urban planning exists in various forms and it addresses many different issues. Urban planning can include urban **renewal**, by adapting urban planning methods to existing cities suffering from decline. **Alternatively**, it can concern the massive challenges associated with urban growth, particularly in the Global South.

In the late 20th century, the term sustainable development has come to represent an ideal **outcome** in the sum of all planning goals.

2 Blueprint planning

Following the rise of empiricism during the industrial revolution, the rational planning movement (1890–1960) emphasized the improvement of the built environment based on key spatial factors. Examples of these factors include: **exposure** to direct sunlight, movement of **vehicular** traffic, standardized housing units, and **proximity** to green–space. To identify and design for these spatial factors, rational planning relied on a small group of highly specialized technicians, including architects, urban designers, and engineers. Other, less common, but nonetheless influential groups included governmental officials, private developers, and landscape architects. Through the strategies associated with these professions, the rational planning movement developed a collection of techniques for quantitative assessment, **predictive** modeling, and design. Due to the high level of training required to grasp these methods, however, rational planning fails to provide an **avenue** for public **participation**. In both theory and practice, this shortcoming opened rational planning to claims of **elitism** and social **insensitivity**.

Although it can be seen as an **extension** of the sort of civic **pragmatism** seen in Oglethorpe's plan for Savannah or William Penn's plan for Philadelphia, the roots of the rational planning movement lie in Britain's **Sanitary** Movement. During this period, **advocates** such as Charles Booth and Ebenezer Howard argued for **central** organized, **top–down** solutions to the problems of industrializing cities. In keeping with the rising power of industry, the source of planning authority in the Sanitary Movement included both traditional governmental offices and private development **corporations**. In London and its surrounding suburbs, **cooperation** between these two entities created a network of new communities clustered around the expanding rail system. Two of the best examples of these communities are Letchworth in Hertfordshire and Hampstead Garden Suburb in Greater London. *In both communities, architects Raymond Unwin and Richard Barry Parker exemplify the* ***elite,*** *top–down approach associated with the rational planning movement by using the planning process to establish a* ***uniform*** *landscape and architectural style based on an idealized* ***medieval*** *village.*

From Britain, the rational planning movement spread out across the world. In areas undergoing industrialization themselves, British influences combined with local movements to create **unique reinterpretations** of the rational planning process. In Paris, architect Le Corbusier adopted rational planning's **centralized** approach and added to it a dedication to **quantitative** assessment and a love for the automobile. Together, these two factors yielded the influential planning aesthetic known as "Tower in the Park". In the United States, Frank Lloyd Wright similarly identified vehicular **mobility** as a principal planning **metric**. However, where Le Corbusier emphasized design through quantitative assessment of spatial processes, Wright identified the insights of local public technicians as the key design criteria. Wright's Broadacre City provides a **vivid** expression of what this landscape might look like.

Throughout both the United States and Europe, the rational planning movement declined in the later half of the 20th century. The reason for the movement's decline was also its strength. By focusing so much on design by technical elites, rational planning lost touch with the public it hoped to serve. Key events in this decline in the United States include the **demolition** of the Pruitt–Igoe housing project in St. Louis and the national **backlash** against urban renewal projects, particularly urban **expressway** projects.

3 Synoptic planning

After the "fall" of blueprint planning in the late 1950s and early 1960s, the synoptic model began to emerge as a dominant force in planning. Lane (2005) describes synoptic planning as having four central elements:

"(1) an **enhanced** emphasis on the specification of goals and targets; (2) an emphasis on quantitative analysis and **predication** of the environment; (3) a concern to identify and evaluate **alternative** policy options; and (4) the evaluation of means against ends."

Public participation was first introduced into this model and it was generally **integrated** into the system process described above. However, the problem was that the idea of a single public interest still dominated attitudes, **effectively devaluing** the importance of participation because it suggests the idea that the public interest is relatively easy to find and only requires the most minimal form of participation.

Blueprint and synoptic planning both **employ** what is called the rational paradigm of planning. The rational model is perhaps the most widely accepted model among planning practitioners and scholars, and is considered by many to be the **orthodox** view of planning. As its name clearly suggests, the goal of the rational model is to make planning as rational and systematic as possible. ***Proponents*** *of this* ***paradigm*** *would generally come up with a list of steps that the planning process can be at least relatively* ***neatly sorted out*** *into and that planning practitioners should go through in order when setting out to plan in virtually any area.* As noted above, this paradigm has clean **implications** for public **involvement** in planning decisions.

4 Participatory planning

Participatory planning is an urban planning paradigm that emphasizes involving the entire community in the strategic and management processes of urban planning; or, a community-level planning process, urban or rural. It is often considered as part of community development. Participatory planning aims to **harmonize** views among all of its participants as well as prevent **conflict** between opposing parties. In addition, **marginalized** groups have an opportunity to **participate** in the planning process.

5 Incremental planning

Beginning in the late 1950s and early 1960s, critiques of the rational paradigm began to emerge and formed into several different schools of planning thought. The first of these schools is Lindblom's **incrementalism**. Lindblom describes planning as "muddling through" and thought that practical planning required decisions to be made **incrementally**. *This incremental approach meant choosing from small number of policy approaches that can only have a small number consequences and are firmly bounded by reality, constantly* ***adjusting*** *the objectives of the planning process and using* ***multiple*** *analyses and evaluations.* Lane (2005) explains the public involvement implications of this philosophy. Though this **perspective** of planning could be considered a large step forward in that it recognizes that there are number of "public interests" and because it provides room for the planning process to be less centralized and **incorporate** the voices other than those of planners, it in practice would only allow for the public to be involved in a minimal, more **reactive** rather than **proactive** way.

6 Mixed scanning model

The mixed scanning model, developed by Etzioni, takes a similar, but slightly different approach. Etzioni (1968) suggested that organizations plan on two different levels: the **tactical** and the **strategic**. He **posited** that organizations could accomplish this by essentially scanning the environment on multiple levels and then choose different strategies and tactics to address what they found there. While Lindblom's approach only operated on the functional level, Etzioni argued, the mixed scanning approach would allow planning organizations to work on both the **functional** and more big-picture **oriented** levels. Lane (2005) explains though, that this model does not do much more at improving public involvement since the planner or planning organization is still at its focus and since its goal is not necessarily to achieve **consensus** or **reconcile** differing points of view on a particular subject.

By the late 1960s and early 1970s, planners began to look for new approaches because it was realized that the current models were not necessarily sufficient. As had happened before, a number of different models emerged. Lane (2005) notes that it is most useful to think of these model as emerging from a social **transformation** planning tradition as opposed to a social guidance one, so the emphasis is more **bottom-up** in nature than it is top-down.

7 Transactive planning

Transactive planning was a radical break from previous models. Instead of considering public participation as method that would be used in addition to the normal training planning process, participation was a central goal. For the first time, the public was encouraged to take on an active role in the policy setting process, while the planner took on the role of a distributor of information and a **feedback** source. Transactive planning focuses on **interpersonal** dialogue that develops ideas, which will be turned into action. One of the central goals is **mutual** learning where the planner gets more information on the community and citizens become more educated about planning issues.

8 Advocacy planning

Formulated in the 1960s by lawyer and planning scholar Paul Davidoff, the advocacy planning model takes the perspective that there are large **inequalities** in the political system and in the bargaining process between groups that result in large numbers of people unorganized and unrepresented in the process. It concerns itself with ensuring that all people are **equally** represented in the planning process by advocating for the interests of the **underprivileged** and seeking social change. Again, public participation is a central **tenet** of this model. A **plurality** of public interests is assumed, and the role of planner is essentially the one as a facilitator who either advocates directly for **underrepresented** groups or directly encourages them to become part of the process.

9 Radical Planning

Radical planning is a stream of urban planning which seeks to manage development in an **equitable** and community-based manner. The **seminal** text to the radical planning movement is *Foundations for a Radical Concept in Planning (1973)*, by Stephen Grabow and Allen Heskin. Grabow and Heskin provided a critique of planning as **elitist**, centralizing and change-resistant, and proposed a new paradigm based upon systems change, **decentralization**, communal society, facilitation of human development and a consideration of **ecology**. Grabow and Heskin were joined by Head of Department of Town Planning from the **Polytechnic** of the South Bank Shean McConnell, and his 1981 work *Theories for Planning*.

In 1987 John Friedman entered the **fray** with *Planning in the Public Domain: From Knowledge to Action*, **promoting** a radical planning model based on "**decolonization**", "**democratization**", "**self-empowerment**" and "**reaching out**". Friedman described this model as an "**agropolitan** development" paradigm, emphasizing the **re-localization** of primary production and manufacture. In *Toward a* ***Non-Euclidian*** *Mode of Planning* (1993) Friedman further promoted the urgency of **decentralizing** planning, advocating a planning paradigm that is **normative, innovative**, political, transactive and based on a social learning approach to knowledge and policy.

10 Bargaining model

The bargaining model views planning as the result of give and take on the part of a number of interests who are all involved in the process. It argues that this bargaining is the best way to conduct planning within the bounds of legal and political **institutions**. The most interesting part of this theory of planning is that it makes public participation the central **dynamic** in the decision-making process. Decisions are made first and foremost by the public, and the planner plays a more minor role.

11 Communicative approach

The communicative approach to planning is perhaps the most difficult to explain. It focuses on using **communication** to help different interests in the process understand each other. The idea is that each individual will approach a conversation with his or her own subjective experience in mind and that from that conversation shared goals and possibilities will emerge. Again, participation plays a central role under this model. The model seeks to include as a broad range of voice to enhance the debate and negotiation that is supposed to form the core of actual plan making. In this model, participation is actually fundamental to the planning process happening. Without the involvement of concerned interests there is no planning. Bent Flyvbjerg and Tim Richardson have developed a critique of the communicative approach and an alternative theory based on an understanding of power and how it works in planning.

Looking at each of these models it becomes clear that participation is not only shaped by the public in a given area or by the attitude of the planning organization or planners that work for it. In fact, public participation is largely influenced by how planning is defined, how planning problems are defined, the kinds of knowledge that planners choose to employ and how the planning **context** is set. Though some might argue that is too difficult to involve the public through transactive, advocacy, bargaining and communicative models because transportation is some ways more technical than other fields, it is important to note that transportation is perhaps unique among planning fields in that its systems depend on the **interaction** of a number of individuals and organizations.

12 Changes to the planning process

Prior to 1950, urban planning was seldom considered a unique profession in Canada. There were, and are, of course, differences from country to country. For example, the UK's Royal Town Planning Institute was created as a professional organization in 1914 and given a Royal Charter in 1959. Town planning focused on top–down processes by which the urban planner created the plans. The planner would know architecture, surveying, or engineering, bringing to the town planning process ideals based on these disciplines. They typically worked for national or local governments. Urban planners were seen as generalists, capable of integrating the work of other disciplines into a **coherent** plan for whole cities or parts of cities. A good example of this kind of planner was Lewis Keeble and his standard textbook, *Principles and Practice of Town and Country Planning*, published in 1951.

Strategic urban planning over past decades have witnessed the **metamorphosis** of the role of the urban planner in the planning process. More citizens calling for **democratic** planning & development processes have played a huge role in allowing the public to make important decisions as part of the planning process. Community organizers and social workers are now very involved in planning from the **grassroots** level. *The term advocacy planning was coined by Paul Davidoff in his influential 1965 paper, Advocacy and* ***Pluralism*** *in Planning which acknowledged the political nature of planning and urged planners to acknowledge that their actions are not* ***value–neutral*** *and encouraged minority and underrepresented voices to be part of planning decisions.* Benveniste argued that planners had a political role to play and had to bend some truth to power if their plans were to be **implemented**.

Developers have also played huge roles in development, particularly by planning projects. Many recent developments were results of large and small-scale developers who purchased land, designed the district and constructed the development **from scratch**. The Melbourne Docklands, for example, was largely an **initiative** pushed by private developers to redevelop the **waterfront** into a **high-end** residential and commercial district.

Recent theories of urban planning, **espoused**, for example by Salingaros see the city as an adaptive system that grows according to process similar to those of plants. They say that urban planning should thus take its **cues** from such natural processes. Such theories also advocate participation by inhabitants in the design of the urban environment, as **opposed** to simply leaving all development to large-scale construction firms.

Notes

[1] 本文源自维基百科（From Wikipedia, the free encyclopedia, https://www.wikipedia.org/）。

[2] 本文主要相关文献：栾峰，2004; 孙施文，2007; 于泓，2003; 张庭伟，1999; 周国艳，2010; Davidoff, 1965; Etzioni, 1967; Friedmann, 1987; Forester, 1993, 1999; Healey, 1988, 1997; Innes, 1995; Lindblom, 1959; Sager, 1994; Taylor, 1998。

[3] 课文中下划线部分为有关城市规划理论的重要论述，需强化阅读，每篇课文习题部分均要求将下划线句子或段落译成中文。

[4] 课文中下划线斜体字部分为复杂长句，其解析详见下文“**Complex Sentences**”。

New Words

adjust v. 适应，调整〖前缀 3. ad-; 词根 102. just〗
advocacy n. 拥护，支持〖前缀 3. ad-; 词根 232. voc〗
advocate v. 拥护，支持，鼓吹；n. 支持者，拥护者〖前缀 3. ad-; 词根 232. voc〗
agropolitan adj. 农业的，农村的
alternative adj. 择一的，供选择的；n. 选择，替换物〖词根 7. alter〗
alternatively adv. 择一地；供选择地〖词根 7. alter〗
avenue n. 林荫路，大街；途径，手段
backlash n. 反击
bargain v. 讨价还价，做交易
blueprint n. 蓝图
bottom-up adj. 自下而上的
central adj. 中心的，中央的〖词根 32. centr〗
centralize v. 使集中，成为……的中心〖词根 32. centr〗

coherent adj. 一致的，连贯的〖前缀 16. co-; 词根 90. here〗

communication n. 通信；交流

communicative adj. 爱说话的，好社交的

conflict n. 冲突，战斗〖前缀 16. con-; 词根 73. flict〗

consensus n. 意见一致〖前缀 16. con-; 词根 186. sens〗

context n. 上下文；背景，环境；语境

cooperation n. 合作，协作〖前缀 16. co-〗

corporation n. 公司，社团〖词根 42. corpor〗

cue n. 暗示，提示，线索

decentralization n. 分散〖前缀 18. de-; 词根 32. centr〗

decentralize v. 分散〖前缀 18. de-; 词根 32. centr〗

decolonization n. 非殖民地化〖前缀 18. de-〗

democratic adj. 民主的〖词根 54. dem; 词根 44. crat〗

democratization n. 民主化〖词根 54. dem; 词根 44. crat〗

demolition n. 毁坏，破坏，拆毁

devalue v. 贬低，使贬值〖前缀 18. de-〗

dynamic n. 动态；动力

ecology n. 生态学〖词根 57. eco〗

effectively adv. 有效地

elite n. 精华，精英，掌权人物

elitism n. 精英统治，精英主义

elitist n. 优秀人才，杰出人物；adj. 优秀人才的，杰出者的

employ v. 雇佣，使用，利用

enhance v. 提高，增加，加强

equally adv. 平等地，公平地〖词根 58. equ〗

equitable adj. 公平的，公正的〖词根 58. equ〗

espouse v. 拥护，赞助

exposure n. 暴露，展出，公开〖前缀 26. ex-; 词根 170. pos〗

expressway n. 高速公路

extension n. 伸展，扩大，延长〖前缀 26. ex-; 词根 208. tens〗

feedback n. 反馈

fray n. 争辩，争吵

from scratch 从头做起，从零开始

functional adj. 功能的〖词根 79. funct〗

grassroot n. 草根，基层，基层民众；adj. 基层的

harmonize v. 和谐

high-end adj. 高档的，高档次的

implement v. 实施，执行

implication n. 暗示，含意；卷入，牵连〖前缀 33. im-; 词根 169. pli〗

incorporate v. 合并，并入〖前缀 34. in-; 词根 42. corpor〗

incremental adj. 增加的〖前缀 34. in-; 词根 45. cre〗

incrementalism n. 渐进主义〖前缀 34. in-; 词根 45. cre〗

incrementally adv. 逐渐地〖前缀 34. in-; 词根 45. cre〗

inequality n. 不平等，不均等〖前缀 34. in-; 词根 58. equ〗

initiative n. 倡议，主动性，主动权

innovative adj. 革新的，创新的〖前缀 34. in-; 词根 141. nov〗

insensitivity n. 不灵敏，不敏感〖前缀 34. in-; 词根 186. sens〗

institution n. 制度，机制；社会公共机构，学校

institutional adj. 制度的，机制的；公共机构的，慈善机构的

institutionalize v. 使制度化

integral adj. 构成整体所必需的，完整的

integrate v. 使成整体

integrative adj. 综合的

integrity n. 完整；正直，诚实

interaction n. 互动，相互影响〖前缀 35. inter-; 词根 2. act〗

interpersonal adj. 人与人之间的，人际的〖前缀 35. inter-〗

involvement v. 牵连，加入〖前缀 34. in-; 词根 234. volv〗

marginalize v. 使处于边缘，忽视

medieval adj. 中世纪的〖词根 122. medi〗

metamorphosis n. 变形〖前缀 39. meta-; 词根 131. morph〗

metric n. 度量标准〖词根 126. metr〗

mobility n. 流动性，移动性〖词根 130. mob〗

muddle v. 弄乱，弄糟；使糊涂

multiple adj. 多重的，多个的，复杂的〖前缀 43. multi-〗

mutual adj. 相互的，彼此的

neatly adv. 整洁地，干净地

non-Euclidian adj. 非欧几里得的，非欧几里得几何学的〖前缀 45. non-〗

normative adj. 标准的，规范的〖词根 138. norm〗

oppose v. 反对，抗争〖前缀 op-, 反；词根 170. pos〗

orient v. 确定方向；使熟悉情况；n. 东方〖词根 148. ori〗

origin n. 起源，根源〖词根 148. ori〗

orthodox adj. 正统的〖词根 149. ortho〗

outcome n. 结果，成果〖前缀 47. out-〗

paradigm n. 范例，样式

participate v. 参加，参与

participation n. 参加，参与

participatory adj. 提供参加机会的，供人分享的

perspective n.（判断事物的）角度，方法；透视法〖前缀 53. per-; 词根 198. spec〗

pluralism n. 多元化，多元性

plurality n. 多元化

polytechnic adj. 有关多种工艺的，工艺教育的〖前缀 55. poly-; 词根 205. tech〗

posit v. 假定，设想

pragmatism n. 实用主义

predication n. 断言，断定

predictive adj. 预言的〖前缀 57. pre-; 词根 55. dict〗

proactive adj. 积极主动的〖前缀 58. pro-; 词根 2. act〗

promote v. 促进，推进，提升〖前缀 58. pro-; 词根 132. mot〗

promotion n. 促进，推进，提升〖前缀 58. pro-; 词根 132. mot〗

promotive adj. 促进的〖前缀 58. pro-; 词根 132. mot〗

proponent n. 支持者，拥护者

proximity n. 接近，邻近，亲近

quantitative adj. 定量的

reach out v. 伸出

reaction n. 反应，反作用力；反动，保守〖词根 2. act〗

reactive adj. 反应的〖词根 2. act〗

reconcile v. 和解，调和

reinterpretation n. 重新解释，纠正性说明〖前缀 61. re-〗

re-localization n. 再定位〖前缀 61. re-〗

renewal n. 重建，更新〖前缀 61. re-〗

sanitary adj. 清洁的，卫生的

scan v. 扫描；细看，细查

self-empowerment n. 自我授权〖前缀 64. self-〗

seminal adj. 创新的，有重大影响的〖词根 185. semin〗

sort out v. 分类，辨别出来

strategic adj. 战略的

synoptic adj. 提纲的，概要的

tactical adj. 战术的

tenet n. 原则，信条，教义

top-down adj. 自上而下的

transact v. 处理，办理；商议，谈判〖前缀 71. trans-; 词根 2. act〗

transformation n. 转化，转变〖前缀 71. trans-; 词根 76. form〗

underprivileged adj. 贫穷的，下层社会的，被剥夺基本权利的〖前缀 75. under-〗

underrepresented adj. 未被充分代表的〖前缀 75. under-〗

uniform adj. 规格一致的，始终如一的〖前缀 76. uni-; 词根 76. form〗

unique adj. 唯一的，独一无二的〖前缀 76. uni-〗

value-neutral adj. 价值观中立的

vehicular adj. 车的，用车辆运载的

vivid adj. 生动的〖词根 231. viv〗

waterfront n. 海滨，江边

Complex Sentences

[1] In both communities, **architects Raymond Unwin and Richard Barry Parker exemplify the elite, top-down approach** associated with the rational planning movement by using the planning process to establish a uniform landscape and architectural style based on an idealized medieval village.

黑体字部分为句子的主干（主语、谓语和宾语），associated 引导的过去分词短语作定语，修饰前面的 the elite, top-down approach，by 引导的介词短语作状语，based 引导的过去分词短语作定语，修饰前面的 a uniform landscape and architectural style。

[2] **Proponents of this paradigm would generally come up with a list of steps** that the planning process can be at least relatively neatly sorted out into and that planning practitioners should go through in order when setting out to plan in virtually any area.

黑体字部分为句子的主干（主语、谓语和宾语），其后两个 that 引导的从句均为定语从句，修饰前面的 a list of steps。

[3] **This incremental approach meant choosing from small number of policy approaches** that can only have a small number consequences and are firmly bounded by reality, constantly adjusting the objectives of the planning process and using multiple analyses and evaluations.

黑体字部分为句子的主干（主语、谓语和宾语），其后 that 引导的从句为定语从句，修饰前面的 small number of policy approaches，adjusting 和 using 引导的现在分词短语共同作状语。

[4] **The term advocacy planning was coined** by Paul Davidoff in his influential 1965 paper, Advocacy and Pluralism in Planning *which **acknowledged** the political nature of planning and **urged** planners to acknowledge that their actions are not value-neutral and **encouraged** minority and underrepresented voices to be part of planning decisions.*

黑体字部分为句子的主干，斜体字部分为 which 引导的定语从句，修饰前面的 paper，即 Advocacy and Pluralism in Planning；该定语从句结构复杂，包括三个谓语：acknowledged, urged 以及 encouraged。

Exercises

[1] 将课文中下划线句子或段落译成中文。

[2] 读熟或背诵上述复杂长句。

[3] 解析下列单词 [解析单词是指根据单词中包含的词根、前缀推导出单词的含义，例如，apathy（n. 漠然，冷淡），解析式为：a-（无）+ path（感情）+ y →无感情→冷漠，冷淡]。解析的单词选自上文"**New Words**"。

centralize v. 使集中，成为……的中心	oppose v. 反对，抗争
conflict n. 冲突，战斗	orthodox adj. 正统的
consensus n. 意见一致	perspective n.（判断事物的）角度，方法；透视法
devalue v. 贬低，使贬值	polytechnic adj. 有关多种工艺的，工艺教育的
exposure n. 暴露，展出，公开	predictive adj. 预言的
incremental adj. 增加的	proactive adj. 积极主动的
inequality n. 不平等，不均等	promote v. 促进，推进，提升
innovative adj. 革新的，创新的	transact v. 处理，办理；商议，谈判
insensitivity n. 不灵敏，不敏感	underrepresented adj. 未被充分代表的
medieval adj. 中世纪的	unique adj. 唯一的，独一无二的
metamorphosis n. 变形	vivid adj. 生动的

Unit 8

Urban Design

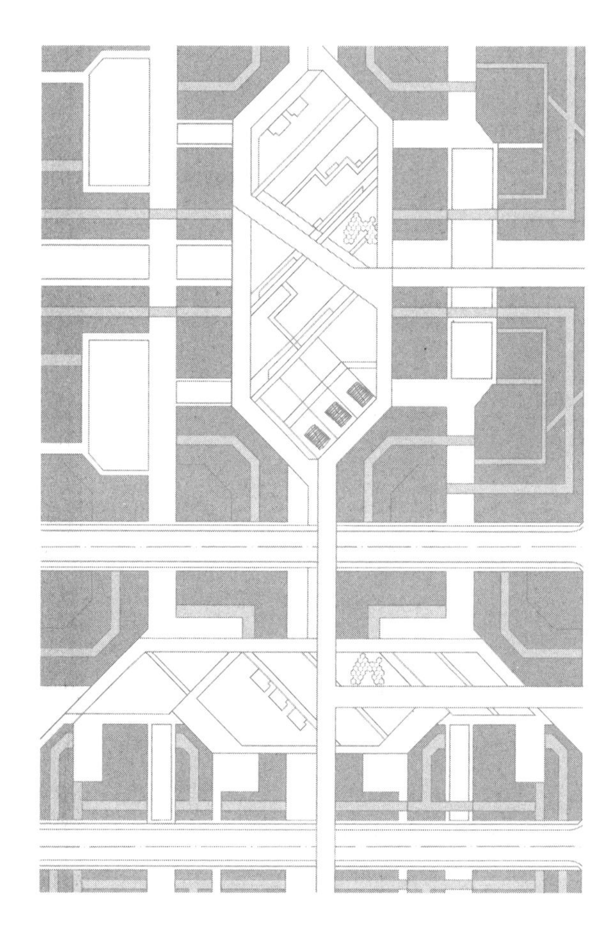

Urban design is the process of designing and shaping cities, towns and villages. *In contrast to architecture, which focuses on the design of individual buildings, urban design deals with the larger scale of groups of buildings, streets and public spaces, whole neighborhoods and districts, and entire cities, with the goal of making urban areas functional, attractive, and sustainable.*

Urban design is an **interdisciplinary** subject that utilizes elements of many built environment professions, including landscape architecture, urban planning, architecture, civil and **municipal** engineering. It is common for professionals in all these **disciplines** to practice in urban design. In more recent times different sub–strands of urban design have emerged such as strategic urban design, landscape urbanism, water–sensitive urban design, and sustainable urbanism.

Urban design demands an understanding of a wide range of subjects from physical geography to social science, and an **appreciation** for disciplines, such as real estate development, urban economics, political economy and social theory.

Urban design is about making connections between people and places, movement and urban form, nature and the built **fabric**. Urban design draws together the many strands of place–making, environmental **stewardship**, social **equity** and economic **viability** into the creation of places with distinct beauty and identity. Urban design draws these and other strands together creating a **vision** for an area and then **deploying** the resources and skills needed to bring the vision to life.

Urban design theory deals primarily with the design and management of public space (i.e. the public environment, public realm or public domain), and the way public places are experienced and used. Public space includes the totality of spaces used freely on a day-to-day basis by the general public, such as streets, plazas, parks and public infrastructure. Some aspects of privately owned spaces, such as building **facades** or domestic gardens, also contribute to public space and are therefore also considered by urban design theory. Important writers on urban design theory include Christopher Alexander, Peter Calthorpe, Gordon Cullen, Andres Duany, Jane Jacobs, Mitchell Joachim, Jan Gehl, Allan B. Jacobs, Kevin Lynch, Aldo Rossi, Colin Rowe, Robert Venturi, William H. Whyte, Camillo Sitte, Bill Hillier and Elizabeth Plater-Zyberk.

1 History

Although **contemporary** professional use of the term "urban design" dates from the mid-20th century, urban design as such has been practiced throughout history. Ancient examples of carefully planned and designed cities exist in Asia, Africa, Europe and the Americas, and are particularly well-known within Classical Chinese, Roman and Greek cultures.

European Medieval cities are often, and often **erroneously**, regarded as exemplars of undesigned or **organic** city development. There are many examples of considered urban design in the Middle Ages. In England, many of the towns listed in the 9th century Burghal Hidage were designed on a **grid**, examples including Southampton, Wareham, Dorset and Wallingford, Oxfordshire, having been rapidly created to provide a **defensive** network against Danish **invaders**. 12th century western Europe brought renewed focus on urbanization as a means of stimulating economic growth and **generating** revenue. The **burgage** system dating from that time and its associated burgage plots brought a form of self-organizing design to medieval towns. **Rectangular** grids were used in the **Bastides** of 13th and 14th century Gascony, and the new towns of England created in the same period.

Throughout history, design of streets and **deliberate configuration** of public spaces with buildings have reflected **contemporaneous** social **norms** or philosophical and religious beliefs. Yet the link between designed urban space and human mind appears to be **bidirectional**. Indeed, the **reverse** impact of urban structure upon human behavior and upon thought is evidenced by both observational study and historical record. There are clear indications of impact through **Renaissance** urban design on the thought of Johannes Kepler and Galileo Galilei. Already René Descartes in his ***Discourse*** *on the Method* had **attested** to the impact Renaissance planned new towns had upon his own thought, and much evidence exists that the Renaissance **streetscape** was also the **perceptual stimulus** that had led to the development of **coordinate geometry**.

The beginnings of modern urban design in Europe are **associated** with the Renaissance but, especially, with the Age of Enlightenment. Spanish colonial cities were often planned, as were some towns settled by other imperial cultures. These sometimes **embodied utopian ambitions** as well as aims for functionality and good governance, as with James Oglethorpe's plan for Savannah, Georgia. In the Baroque period the design approaches developed in French formal gardens such as Versailles were extended into urban development and **redevelopment**. In this period, when modern professional specializations did not exist, urban design was undertaken by people with skills in areas as **diverse** as sculpture, architecture, garden design, surveying, **astronomy**, and military engineering. In the 18th and 19th centuries, urban design was perhaps most closely linked with surveyors (engineers) and architects. The increase in urban populations brought with it problems of **epidemic** disease, the response to which was a focus on public health, the rise in the UK of municipal engineering and the inclusion in British legislation of **provisions** such as minimum widths of street in relation to heights of buildings in order to ensure **adequate** light and **ventilation**.

Much of Frederick Law Olmsted's work was concerned with urban design, and the newly formed profession of landscape architecture also began to play a significant role in the late 19th century.

1.1 Modern urban design

Urban planning focuses on public health and urban design. Within the discipline, modern urban design developed.

At the turn of the 20th century, planning and architecture underwent a **paradigm** shift because of societal pressures. During this time, cities were industrializing at a **tremendous** rate; private business largely **dictated** the pace and style of this development. The expansion created many hardships for the working poor and concern for health and safety increased. However, the **laissez-faire** style of government, in fashion for most of the Victorian era, was starting to give way to a New **Liberalism**. This gave more power to the public. The public wanted the government to provide citizens, especially factory workers, with healthier environments. Around 1900, modern urban design **emerged** from developing theories on how to **mitigate** the consequences of the industrial age.

The first modern urban planning theorist was Sir Ebenezer Howard. His ideas, although utopian, were adopted around the world because they were highly practical. He initiated the garden city movement in 1898. His garden cities were intended to be planned,

self-contained communities surrounded by parks. Howard wanted the cities to be **proportional** with separate areas of residences, industry, and agriculture. **Inspired** by the Utopian novel *Looking Backward* and Henry George's work *Progress and Poverty*, Howard published his book *Garden Cities of Tomorrow* in 1898. His work is commonly regarded as the most important book in the history of urban planning. He **envisioned** the **self-sufficient** garden city to house 32,000 people on a site 6,000 acres (2,428 hm^2). He planned on a **concentric** pattern with open spaces, public parks, and six **radial boulevards**, 120 ft (37 m) wide, **extending** from the center. When it reached full population, Howard wanted another garden city to be developed nearby. He **envisaged** a cluster of several garden cities as satellites of a central city of 50,000 people, linked by road and rail. His model for a garden city was first created at Letchworth and Welwyn Garden City in Hertfordshire. Howard's movement was extended by Sir Frederic Osborn to regional planning.

In the early 1900s, urban planning became professionalized. With input from utopian **visionaries**, infrastructure engineers, and local councilors, new urban planning **templates** were made to be considered by politicians. In 1899, the Town and Country Planning Association was founded. In 1909, the first academic course on urban planning was offered by the University of Liverpool. Urban planning was first officially embodied in the Housing and Town Planning Act of 1909. Howard's "garden city" **compelled** local authorities to introduce a system where all housing construction **conformed** to specific building standards. In the United Kingdom following this Act, surveyor, civil engineers, architects, and lawyers began working together within local authorities. In 1910, Thomas Adams became the first Town Planning **Inspector** at the Local Government Board and began meeting with practitioners. In 1914, The Town Planning Institute was established. The first urban planning course in America wasn't established until 1924 at Harvard University. Professionals developed schemes for the development of land, **transforming** town planning into a new area of expertise.

In the 20th century, urban planning was forever changed by the automobile industry. Car **oriented** design impacted the rise of "urban design". City layouts now had to **revolve** around roadways and traffic patterns. In 1956, "urban design" was first used at a series of conferences Harvard University. The event provided a platform for Harvard's Urban Design program. The program also utilized the writings of famous urban planning thinkers: Gordon Cullen, Jane Jacobs, Kevin Lynch, and Christopher Alexander.

In 1961, Gordon Cullen published *The **Concise** Townscape*. He examined the traditional artistic approach to city design of theorists including Camillo Sitte, Barry Parker and Raymond Unwin. Cullen also created the concept of "serial vision". It defined the urban landscape as a series of related spaces.

In 1961, Jane Jacobs published *The Death and Life of Great American Cities*. She critiqued the Modernism of CIAM (Congrès internationaux d'architecture moderne, International Congresses of Modern Architecture). Jacobs also considered crime rates in publicly owned spaces were rising because of the Modernist approach of "city in the park". She argued instead for an "eyes on the street" approach to town planning through the **resurrection** of main public space **precedents** (e.g. streets, squares).

In the same year, Kevin Lynch published *The Image of the City*. He was **seminal** to urban design, particularly with regards to the concept of **legibility**. He reduced urban design theory to five basic elements: paths, districts, edges, nodes, landmarks. He also made the use of mental maps to understanding the city popular, rather than the two-dimensional physical master plans of the previous 50 years.

Other notable works:

· *Architecture of the City* by Rossi (1966).

· *Learning from Las Vegas* by Venturi (1972).

· *Collage City* by Colin Rowe (1978).

· *The Next American Metropolis* by Peter Calthorpe (1993).

· *The Social Logic of Space* by Bill Hillier and Julienne Hanson (1984).

The popularity of these works resulted in terms that become everyday language in the field of urban planning. Aldo Rossi introduced "historicism" and "collective memory" to urban design. Rossi also proposed a "**collage metaphor**" to understand the collection of new and old forms within the same urban space. Peter Calthorpe developed a **manifesto** for sustainable urban living via medium density living. He also designed a **manual** for building new settlements in his concept of "**Transit Oriented** Development (TOD)". Bill Hillier and Julienne Hanson introduced "Space **Syntax**" to predict how movement patterns in cities would contribute to urban **vitality**, **anti-social** behavior and economic success. "Sustainability", "livability", and "high quality of urban components" also became **commonplace** in the field.

1.1.1 Current trends

Urban design seeks to create sustainable urban environments with long-lasting structures, buildings and overall livability. Walkable urbanism is another ideal that is defined as the Charter of New Urbanism. It aims to reduce environmental impacts by altering the built environment to create smart cities that support sustainable transport. **Compact** urban neighborhoods encourage residents to drive less. These neighborhoods have significantly lower environmental impacts when compared to **sprawling** suburbs. To prevent urban sprawl, **Circular** flow land use management was introduced in Europe to

promote sustainable land use patterns.

As a result of the recent New Classical Architecture movement, sustainable construction aims to develop smart growth, walkability, architectural tradition, and classical design. It contrasts from modernist and globally **uniform** architecture. In the 1980s, urban design began to oppose the increasing **solitary** housing estates and suburban sprawl.

2 Principles

Public agencies, authorities, and the interests of nearby property owners manage public spaces. Users often compete over the spaces and negotiate across a **variety** of spheres. Urban designers lack control in the design profession; this is offered to architects. Input from engineers, **ecologists**, local historians, and transportation planners is required to offer a balanced representation of ideas.

Urban designers are similar to urban planners when preparing design guidelines, regulatory frameworks, legislation, advertising, etc. Urban planners also overlap with architects, landscape architects, transportation engineers and industrial designers. They must also deal with "place management" to guide and assist the use and maintenance of urban areas and public spaces.

There are professionals who identify themselves specifically as urban designers. However, architecture, landscape and planning programs **incorporate** urban design theory and design subjects into their curricula. There are an increasing number of university programs offering degrees in urban design at post-graduate level.

Urban design considers:

- **Pedestrian** zones.
- **Incorporation** of nature within a city.
- Aesthetics.
- Urban structure–arrangement and relation of business and people.
- Urban **typology**, density and sustainability–spatial types and **morphologies** related to intensity of use, consumption of resources and production and maintenance of viable communities.
- **Accessibility**–safe and easy transportation.
- **Legibility** and wayfinding–accessible information about travel and destinations.
- **Animation**–designing places to stimulate public activity.
- Function and fit–places support their varied intended uses.
- **Complementary** mixed uses–locating activities to allow constructive interaction

between them.

· Character and meaning-recognizing differences between places.

· Order and incident-balancing consistency and variety in the urban environment.

· Continuity and change-locating people in time and place, respecting heritage and contemporary culture.

· Civil society-people are free to interact as civic equals, important for building social capital.

3 Equality issues

Until the 1970s, the design of towns and cities took little account of the needs of people with **disabilities**. At that time, **disabled** people began to form movements demanding recognition of their potential contribution if social obstacles were removed. Disabled people challenged the "medical model" of disability which saw physical and mental problems as an individual "tragedy" and people with disabilities as "brave" for enduring them. They proposed instead a "social model" which said that barriers to disabled people result from the design of the built environment and attitudes of able-bodied people. "**Access** Groups" were established **composed** of people with disabilities who **audited** their local areas, checked planning applications and made representations for improvements. The new profession of "access officer" was established around that time to produce guidelines based on the recommendations of access groups and to **oversee** adaptations to existing buildings as well as to check on the **accessibility** of new proposals. Many local authorities now employ access officers who are regulated by the Access Association. A new chapter of *the Building Regulations (Part M)* was introduced in 1992. Although it was **beneficial** to have legislation on this issue the requirements were fairly minimal but continue to be improved with ongoing **amendments**. *The Disability Discrimination Act 1995* continues to raise awareness and enforce action on disability issues in the urban environment.

Notes

[1] 本文源自维基百科 (From Wikipedia, the free encyclopedia, https://www.wikipedia.org/)。

[2] 本文主要相关文献：Alexander, 1977; Bacon, 1967; Jacobs, et al, 1987; Jacobs, 1995; Lynch, 1961, 1984, 1988; Macdonald, et al, 2007; McHarg, 1969; Sitte, 1889。

[3] 课文中下划线部分为有关城市规划理论的重要论述，需强化阅读，每篇课文习题部分均要求将下划线句子或段落译成中文。

[4] 课文中下划线斜体字部分为复杂长句，其解析详见下文“**Complex Sentences**”。

New Words

access n. 接近，进入，通路；v. 接近，使用〖前缀 3. ac–; 词根 30. cess〗
accessibility n. 可达性〖前缀 3. ac–; 词根 30. cess〗
adequate adj. 足够的；适当的
ambition n. 抱负，雄心，野心〖前缀 5. anbi–〗
amendment n. 修改，修订，修正案〖词根 124. mend〗
animation n. 生气，活泼；动画片〖词根 10. anim〗
anti–social adj. 反社会的〖前缀 7. anti–〗
appreciation n. 欣赏，鉴赏
associate v.（使）发生联系，（使）联合〖词根 193. soci〗
astronomy n. 天文学〖词根 16. astr〗
attest v. 证明，证实
audit v. 审计，查账
bastide n. 中世纪专为防御而建的乡村（城镇）
beneficial adj. 有利的，有益的〖前缀 11. bene–; 词根 65. fic〗
bidirectional adj. 双向的〖前缀 12. bi–〗
boulevard n. 大马路，林荫大道
burgage n. 租地权，土地保有权
circular adj. 圆形的，环形的，循环的〖前缀 15. circu–〗
collage n. 拼贴画，大杂烩
commonplace adj. 平凡的，普通的
compact adj. 紧凑的，简洁的
compel v. 强迫，迫使〖前缀 16. com–; 词根 157. pel〗
complementary adj. 补充的，互补的
compose v. 组成；创作；使平静〖前缀 16. com–; 词根 170. pos〗
concentric adj. 同一中心的〖前缀 16. con–; 词根 32. centr〗
concise adj. 简明的，简洁的〖前缀 16. con–; 词根 35. cise〗
configuration n. 布局，构造，配置〖前缀 16. con–; 词根 67. figure〗
conform v. 符合，遵照〖前缀 16. con–; 词根 76. form〗
contemporaneous adj. 同时期的，同时代的〖前缀 16. con–; 词根 207. tempor〗
contemporary adj. 当代的，同时代的〖前缀 16. con–; 词根 207. tempor〗
coordinate v. 使协调，使调和〖前缀 16. co–; 词根 147. ordin〗

defensive adj. 防御用的，防守的

deliberate adj. 故意的，深思熟虑的〖前缀 18. de-; 词根 108. liber〗

deploy v. 施展，有效地利用

dictate v. 命令，指示；控制，支配〖词根 55. dict〗

disability n. 残疾；无力，无能〖前缀 20. dis-〗

disabled adj. 残疾的，有缺陷的〖前缀 20. dis-〗

discipline n. 学科；纪律

discourse n. 论述，论文

diverse adj. 不同的，多种多样的〖前缀 19. di-; 词根 226. vers〗

ecologist n. 生态学者〖词根 57. eco〗

embody v. 表现，体现〖前缀 23. em-〗

emerge v. 出现，浮现

envisage v. 想象，设想〖前缀 24. en-; 词根 229. vis〗

envision v. 想象，预见，展望〖前缀 24. en-; 词根 229. vis〗

epidemic adj. 流行的〖词根 54. dem〗

equity n. 公平，公道〖词根 58. equ〗

erroneously adv. 错误地，不正确地〖词根 59. err〗

extend v. 延伸，扩大，推广〖前缀 26. ex-; 词根 208. tend〗

fabric n. 构造，结构

facade n. 建筑物的正面，外表〖词根 62. face〗

generate v. 形成，造成，产生〖词根 81. gene〗

geometry n. 几何学〖词根 geo, 地；词根 126. metr〗

grid n. 格子，方格

incorporate v. 包含，吸收，合并〖前缀 34. in-; 词根 42. corpor〗

incorporation n. 合并，编入〖前缀 34. in-; 词根 42. corpor〗

inspector n. 检察员，检阅官〖前缀 34. in-; 词根 198. spec〗

inspire v. 鼓舞，激励，启迪〖前缀 34. in-; 词根 200. spir〗

interdisciplinary adj. 跨学科的〖前缀 35. inter-〗

invader n. 侵略者，侵犯者〖前缀 34. in-; 词根 219. vad〗

laissez-faire adj. 自由放任的；n. 自由放任主义

legibility n. 易读性，易变认

liberalism n. 自由主义〖词根 108. liber〗

manifesto n. 宣言，声明，告示

metaphor n. 隐喻，暗喻〖前缀 39. meta-〗

mitigate v. 缓和，减轻

morphology n. **形态学**〖词根 131. morph〗

municipal adj. **市的，市政的**〖词根 134. mun〗

norm n. **标准，规则**〖词根 138. norm〗

organic adj. 有机的，自然发展的

orient v. **标定方向，使……向东方；adj. 东方的，新生的；n. 东方**〖词根 148. ori〗

oversee v. **监督，监视**〖前缀 48. over-〗

paradigm n. 范例，样式

pedestrian n. **步行者，行人；adj. 徒步的**〖词根 156. ped〗

perceptual adj. 知觉的，感觉的

precedent n. **前例，先例**〖前缀 57. pre-; 词根 30. cede〗

promote v. **促进，推进，提升**〖前缀 58. pro-; 词根 132. mot〗

proportional adj. 成比例的，相称的

provision n. 规定，条款

radial adj. 辐射状的，放射式的

rectangular adj. **矩形的，成直角的**〖词根 177. rect〗

redevelopment n. **再开发，再发展**〖前缀 61. re-〗

Renaissance n. 文艺复兴

resurrection n. **复活，复兴**〖前缀 61. re-; 词根 177. rect〗

reverse adj. **反面的，颠倒的**〖前缀 61. re-; 词根 226. vers〗

revolve v. **使旋转**〖词根 234. volv〗

self-contained adj. **独立的，自给自足的**〖前缀 64. self-〗

self-sufficient adj. **自足的，极为自负的**〖前缀 64. self-〗

seminal adj. **种子的，创新的，有重大影响的**〖词根 185. semin〗

solitary adj. **独自的，独立的**〖词根 194. sol〗

sprawl v. 蔓延，杂乱无章地拓展

stewardship n. 管理工作

stimulus n. 刺激，刺激物，刺激因素

streetscape n. 街景，街景画

syntax n. 句法，语法

template n. 样板，模板

transform v. **改变，改观**〖前缀 71. trans-; 词根 76. form〗

transit n. **运输，运输路线，公共交通系统**〖前缀 71. trans-〗

tremendous adj. 极大的，巨大的

typology n. 类型学

uniform adj. **一样的，规格一致的**〖前缀 76. uni-; 76. 词根 form〗

utopian adj. 乌托邦的，空想的，不切实际的

variety n. 多样，变化〖词根 221. vari〗

ventilation n. 通风，通风设备

viability n. 生存能力，生活力〖词根 227. vi〗

vision n. 视觉，美景〖词根 229. vis〗

visionary n. 空想家，预言家〖词根 229. vis〗

vitality n. 活力，生命力〖词根 230. vit〗

Complex Sentences

[1] In contrast to architecture, which focuses on the design of individual buildings, **urban design deals with the larger scale of groups of buildings, streets and public spaces, whole neighborhoods and districts, and entire cities,** with the goal of making urban areas functional, attractive, and sustainable.

黑体字部分为句子的主干，In contrast to architecture 作状语，which 引导的定语从句修饰前面的 architecture，with 引导的介词短语作状语。

Exercises

[1] 将课文中下划线句子或段落译成中文。

[2] 读熟或背诵上述复杂长句。

[3] 解析下列单词［解析单词是指根据单词中包含的词根、前缀推导出单词的含义，例如，apathy（n. 漠然，冷淡），解析式为：a-（无）+ path（感情）+ y →无感情→冷漠，冷淡］。解析的单词选自上文“**New Words**”。

animation n. 生气，活泼；动画片	envisage v. 想象，设想
anti-social adj. 反社会的	interdisciplinary adj. 跨学科的
bidirectional adj. 双向的	invader n. 侵略者，侵犯者
circular adj. 圆形的，环形的，循环的	morphology n. 形态学
concise adj. 简明的，简洁的	rectangular adj. 矩形的，成直角的
configuration n. 布局，构造，配置	resurrection n. 复活，复兴
conform v. 符合，遵照	reverse adj. 反面的，颠倒的
contemporaneous adj. 同时期的，同时代的	solitary adj. 独自的，独立的
coordinate v. 使协调，使调和	transit n. 运输，运输路线，公共交通系统
diverse adj. 不同的，多种多样的	viability n. 生存能力，生活力

Unit 9

The Systems and Rational Process Views of Planning (1)

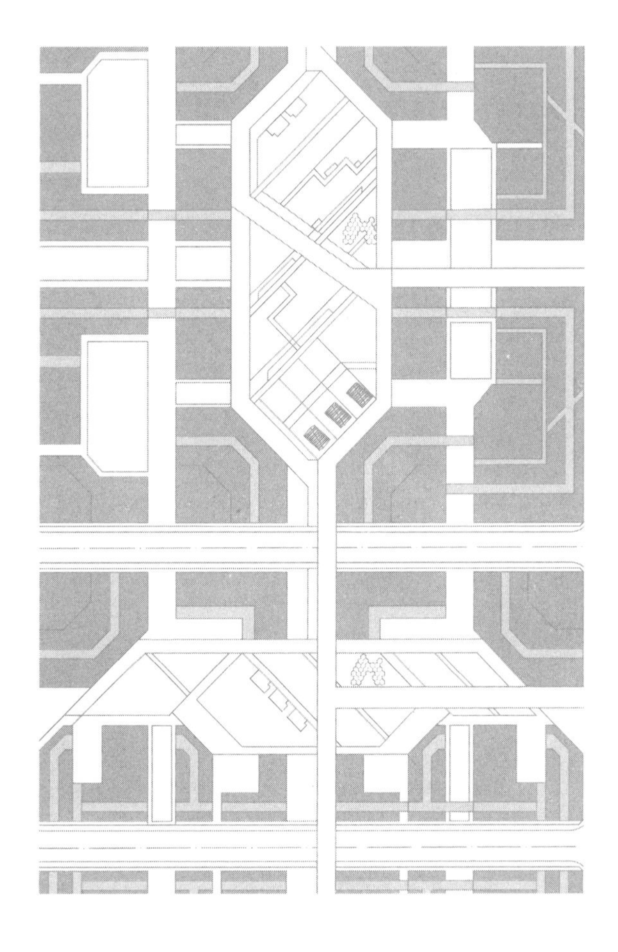

1 Introduction: the radical change in town planning thought

In 1969 a fourth edition of Lewis Keeble's book, *Principles and Practice of Town and Country Planning*, was published. This was still **substantially** the same text as had **originally** been published in 1952. In the same year, 1969, another book was published which, like Keeble's book before it, was to become a standard text for planning students: Brian McLoughlin's *Urban and Regional Planning*: *A Systems Approach*. Figure 1 shows the covers of the 1969 edition of Keeble's book and McLoughlin's book, side by side. Nothing more **vividly captures** the radical change in town planning thought which took place in the 1960s than the contrasting images on these two covers.

On the cover of Keeble's book is one of the author's designs for a **hypothetical** town centre, an image which shows town planning as an exercise in physical planning and town design. Though hypothetical, it **conforms** to established conventions of what a plan for a real town centre might be like; indeed, the diagram could have been the plan of an existing town. The diagram on McLoughlin's book is obviously more abstract. **None the less**, this also **purports** to be a representation of a possible town or city in which the circles and **triangles** represent activities (or land uses) at particular locations, and the lines represent connections between these activities. The varying thicknesses of these lines represent different degrees of **interconnectedness** (e.g. in terms of flows of goods, or people, or traffic, etc.). In other words, the diagram on McLoughlin's book represents an image of the city as an active functioning thing–as a "system".

Why is the cover of McLoughlin's book like this? It is because McLoughlin puts forward the view that town planning is an exercise in the analysis and control of urban areas and regions viewed as systems. So if McLoughlin had been called upon to define town planning, he would have defined it is an exercise in systems analysis and control. This way of **conceiving** planning was a far cry from the post-war Keeblean view of physical planning as an exercise in design.

During the mid to late 1960s, two distinct theories **emerged** which are not (and were not in the 1960s) always clearly distinguished from each other. *One was the "systems view" of planning noted above, which was essentially **derived** from a theory of the object that town planning seeks to plan, namely, the environment (towns, cities, regions, etc.), now seen as a system of interconnected parts.* The other was the "rational process" view of planning, which was a theory about the process of planning and, in particular, of planning as a rational process of decision-making.

Both theories presumed a "deep" conception of planning and control which sociologist Anthony Giddens (1994) has described as a "**cybernetic**" model of planning and politics, and indeed systems theory had direct links with what was called the "science of cybernetics". The idea of cybernetic control has also been associated by Giddens and others with modes of thinking and action characteristic of "modernism". I suggest that the systems and rational process theories of planning, taken together, represented a kind of high **water-mark** of modernist optimism in the post-war era, and in this they shared something in common with the earlier post-war view of planning, in spite of the other differences between these two periods of planning thought.

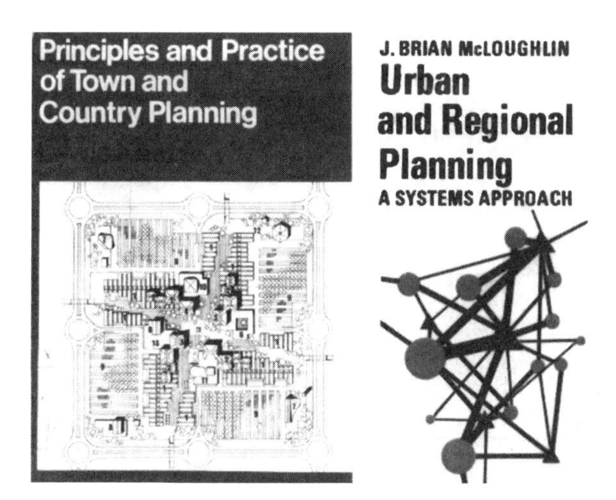

Figure 1 The covers of Lewis Keeble's *Principles and Practice of Town and Country Planning* (1969 edition) and Brian McLoughlin's *Urban and Regional Planning: A System Approach* (1969)

In both this and the following chapter I describe what I term "new planning ideas of the 1960s". This **designation** isn't **arbitrary**; it is fair to say that, in Britain, it was in the 1960s that the ideas I discuss here first had a significant impact on town planning thought. However, these ideas did not just emerge from nowhere, suddenly in the 1960s. Systems theory and rational decision–making **evolved** in other disciplines during the 1940s and 1950s, and they were "**imported**" into town planning in the 1960s. Arguably, it was in the first half of the 1970s that these ideas had their widest influence on planning thought. Thus it was not until 1971 that a "companion" volume to McLoughlin's book was published in Britain, George Chadwick's *A Systems View of Planning*. And it was in 1973 that, again arguably, the leading theorist in Britain of the rational process view of planning, Andreas Faludi, published his influential books *Planning Theory* and *A Reader in Planning Theory*.

2 The systems view of planning

2.1 Basic concepts and their application to planning

The systems view of planning was frequently described in highly abstract, technical and mathematical terms, but the basic ideas of systems theory are really very simple. At the heart of general systems theory is, obviously, the idea of things as "systems". A "system" is something **composed** of interconnected parts. *The Oxford English Dictionary* defines a system as "a complex whole, a set of connected things or parts", and also as "a set or **assemblage** of things connected, associated, or interdependent, so as to form a complex unity". There are two things to note here. First, any system has some kind of **coherence** or **unity** which enables us to distinguish it from other systems and so view it as an entity ("a complex whole") for the purposes of investigation. Thus a system is **analogous** to a "set" in the way this term is used in mathematics, in which what is common to a set is what unifies it and at the same time distinguishes it from other sets. Secondly, what makes a system is not just a set of distinct parts but the fact that the parts are interconnected, and so interdependent. The structure of a system is therefore determined by the structure of its parts and their relationships (see again the diagram on the cover of McLoughlin's book in Figure 1).

The interconnections between the parts of a system are central to its functioning. Consider, for example, human beings as systems. Our bodies **comprise** various **distinguishable** parts: the heart, the lungs, the kidneys, the liver, the brain, etc. When functioning healthily, these parts are actively interconnected via the **circulation** system, which takes blood from the heart to the brain, liver and kidneys, etc. In the technical language of systems theory, one could say that a dead body is one in which the interconnections between the body's parts have "ceased to function".

All living organisms could be viewed as "systems", but the use of the term needs to be qualified here, for any organism depends on, and is therefore related to, its environment. The whole of reality is one **integrated** system, and any system we distinguish within this, such as a living organism, is really a "subsystem" within this larger whole. Because of this, the functioning of any system (or **subsystem**) has to be understood in terms of the ways its parts are "**externally**" interconnected with "parts" of other systems as well as "**internally**" with each other. To take our example of human beings again, the healthy functioning of a person's lungs depends on the quality of the air breathed in from the "external" environment or "**ecosystem**", because any human being is a subsystem within a larger "ecosystem".

Just as we can think of living organisms as systems, so too we can view functioning human–made entities, such as cities and their regions, as systems. A city can be viewed as a system in which its parts are different land–use activities interconnected via transport and other communications media, i.e. a land–use / transport system. As the planning theorist Brian McLoughlin (1965) put it in his earliest article on the systems view of planning, "The components of the urban system are land uses and locations which **interact** through and with the communications networks". This only describes the objects town planning deals with, but the systems view of planning follows logically from this concept of the environment as a system. If the physical environments (towns, cities, regions, etc.) town planning seeks to plan and control are systems, it follows that town planning can be defined as a form of systems control. Or, to put this more fully, since exercising intelligent control over a system requires a prior understanding of the system to be controlled, then we can define town planning as a form of systems analysis and control.

This way of seeing town planning was not entirely new. In the early years of the twentieth century, the pioneering planner Patrick Geddes wrote of cities and their regions as functioning entities, **analogous** in this respect to living organisms (Geddes, 1915; interestingly, Geddes was trained as a botanist). *However, apart from his **strictures** on the need to undertake surveys prior to planning (see below), Geddes's ideas remained **marginal** to the mainstream of town planning thought throughout the first half of the twentieth century, which continued to be dominated by architectural ideas.* And so by the 1960s, against the background of a design–based view of planning, the systems view struck many planners as **novel**, even revolutionary. Below, I describe five major differences between the systems view and the traditional design–based view described in Chapter 1.

Once it was acknowledged that cities (or regions, etc.) were complex systems, it became all the more clear that planners needed to understand "how cities worked". Geddes had emphasized the importance of undertaking surveys prior to the preparation of plans, and his method of "survey–analysis–plan" had been widely adopted. Yet town planners had not acquired a deep understanding of how cities actually functioned. As noted in Chapter 3, Jacobs's and Alexander's main criticism of traditional planning theory was that it showed a serious lack of understanding of the complex reality which planners were dealing with. Alexander drew attention to the fact that planners seemed to lack an appreciation of the complex and rich inter–relationships between phenomena which give rise to successful cities.

Secondly, once cities are viewed as inter–related systems of activities and places, it follows that a change to one part of the city will cause changes to some other part. Any proposed new development must be evaluated in terms of its probable effects, including its effects on activities and places far beyond the actual sites where the new development was proposed. For example, in considering a proposal for a major new shopping development on the edge of a town, planners should also examine the likely effects of this development on, for example, the town centre shops, the local transport system, the likelihood of other development pressures arising in its wake and then of the further effects of that new development and so on.

This was a significantly different way of examining and assessing development proposals from that which had been typically undertaken by planners who viewed planning largely in terms of design and aesthetics. It suggested the need for a new kind of planner altogether, one who was trained in analyzing and understanding how cities and regions functioned spatially in economic and social terms—a planner, that is, trained in economic geography or the social sciences rather than architecture or surveying. Hence McLoughlin's suggestion that the appropriate theoretical understanding needed by town planners was to be found in "location theory", and he devoted a whole chapter of his book to introduce this theory to planners.

Thirdly, as noted in Chapter 3, there were serious questions about whether it was appropriate to produce detailed "end–state" master plans. Systems theory, with its emphasis on activity, **dynamism** and change, suggested the need for more adaptable **flexible** plans—such as the **broad–brush** "structure" plans proposed in the PAG report. When McLoughlin described "plan formulation", he **envisaged** town plans as "trajectories", not end–state blueprints for a fixed future. As he put it: "The form of the plan is that of a trajectory of states at suitable time–intervals".

Fourthly, acceptance of urban change also suggested a view o f town planning as an ongoing process of monitoring, analyzing and intervening in **fluid** situations, rather than an exercise in producing "**once–and–for–all**" blueprints for the ideal future form of a town or city.

Fifthly, viewing cities (or other **discrete** areas of the environment) as systems of interconnected activities implied considering them economically and socially, not just physically and aesthetically. This suggested a much broader and more ambitious **remit** for planning than previously (remember Keeble's **dictum** that town and country planning was not economic and social planning).

This conception of planning was **illustrated** in *The Spirit and Purpose of Planning* (Bruton, 1974) where chapters entitled "Social planning" and "Economic planning" sit alongside one entitled "Physical planning". This broader concept of planning was reflected in the new "structure" plans introduced by *the Town and Country Planning Act 1968*. These plans were specifically intended to be strategic planning documents and their purposes were considered to be as much economic and social as physical. Planning was no longer defined as chiefly involving skills of design and physical planning (though both these more traditional concerns would have been acknowledged as having a more significant place in detailed local planning work and development control).

Related to this last point, in the late 1960s a gap developed between planning theory and the practice of town planning at the local level. No doubt this was **attributable** to the abstract, highly technical (and frankly abstruse) language of systems theory, with its talk of mathematical modeling, "optimization" and so on. The division also **derived** from the fact that much of the work of local authority planning offices continued to be at the level of "local" planning and development control where a constant stream of applications for planning permission had to be dealt with. At this level matters of design and aesthetics continued to be regarded as central. Planning theory concerned with much broader systemic considerations tended to be seen as **irrelevant** by the everyday local planner with a heavy case-load.

It was therefore only in the newly emergent field of strategic, "structure" planning introduced by *the Town and Country Planning Act 1968*, that some of the ideas associated with systems theory were considered by practitioners as relevant to their practice. Keeble's book had been much used in planning offices, and whatever one thinks of that book as a theoretical work, through it the worlds of planning theory and practice met. The same could not be said of McLoughlin's *Urban and Regional Planning: A Systems Approach*, still less of Chadwick's later *A Systems View of Planning*, in which certain passages were specially **marked off** for the "faint-hearted" planner who was "less than **literate** mathematically" (Chadwick, 1971)!

2.2 The rise of the systems view of planning

The emergence of the systems view of planning could be explained as a response to the criticisms of the traditional "physical design" view of town planning. The systems view of planning certainly seemed to meet three of those criticisms. In **concentrating** on the physical and aesthetic qualities of the environment, traditional town planning theory and practice exhibited a lack of understanding of the social and economic life of cities. With its aim of understanding how cities worked as activity systems, the systems view implied a **commitment** to understanding the social and economic life of cities. Secondly, traditional town planning theory exhibited a lack of understanding of the complexity and inter-relatedness of urban life. With its **avowed** aim of seeking to analyze and "model" the complex inter-relationships of cities as systems, the systems view promised to meet this criticism. Finally, the traditional "master" plans had been criticized for their **inflexibility**. With its stress on strategic and **flexible** plans which were sensitive to the dynamic, changing nature of cities, the systems view promised to overcome this problem too.

The emergence of the systems view of planning can thus be seen as a logical response to the **deficiencies** of the physicalist theory. However, it would be **naive** to suppose that developments in planning thought occur in this logical way or that they occur simply as a result of developments within planning thought. Wider forces were at work that contributed to the rise of this new theory. As the American historian of science Thomas Kuhn (1962) has shown, fundamental changes in thought (what he calls "paradigm shifts") are not just driven by the **accumulation** of evidence and the rational response to this evidence. Wider technological, sociological and psychological factors also play a part.

The inter-relatedness of urban phenomena, and specifically land use and transport, had been widely recognized by transport planners in both the USA and the UK, and had already been **highlighted** by the 1963 Buchanan report. This, coupled with rising car use and traffic problems, generated calls for proper "land-use / transportation" studies. Transport planners in the 1960s were mostly traffic engineers, and mathematics was central to their training. Mathematically "modeling" land-use / transport relationships and flows was thus something transport planners took up with enthusiasm. Systems theory had **originated** in the highly technical fields of operations research ("OR") and **cybernetics** where the precise modeling of systemic relationships using statistical and mathematical techniques was seen as necessary to control systems. The development of computers capable of handling the data derived from **numerous** systemic relationships also encouraged the use of systems theory in town and country planning.

Systems theory had an impact on a number of other disciplines. In academic geography, for example, the traditional concern with space and location was translated into viewing settlements and land uses as locations within networks of inter-related places—as spatial systems (see, e.g. Haggett, 1965). As the town planning profession had been opened up to graduates from disciplines other than architecture and surveying, etc., and as it was geographers who mostly exploited this opening, **cross-fertilization** occurred between these two disciplines. As noted earlier, a whole chapter of McLoughlin's book (1969) promoted geographical work on location theory rather than design theory, as the necessary theoretical foundation for planning.

The "quantitative revolution" in 1960s geography was driven by a desire to make geographical studies more precise and "scientific" rather than an "art". *The **impressively** sounding language of systems theory and analysis and rigorous statistical methods of investigation promised to place geography on a much firmer theoretical and scientific foundation, thereby improving its standing as an academic discipline.* The same could be said of planning, which had also been traditionally conceived as an art. The **trenchant** criticisms of writers such as Jane Jacobs seemed to show that the practice of planning lacked any firm theoretical foundation. Systems theory, with its claim to be "scientific", seemed to offer this hope for planning, just as it did for geography.

Finally, the **ecological** thinking of the late 1960s emphasized the interrelatedness of natural phenomena in "ecosystems". Again, the opening chapter of McLoughlin's book, introduced the reader to the basic ideas of systems thinking by describing "man in his ecological setting", and by **illustrating** the practical relevance of understanding systemic relationships through examples of human actions which had **irreparably** damaged natural ecosystems.

*So although the emergence of the systems view of planning in the 1960s can be explained in part as a rational response to the **alleged deficiencies** of the traditional design-based theory of planning, this is only part of the story.* In the case of planning, following the **damning** criticisms which had been made of its practice and theory in the early 1960s, as much as anything there was a felt need for the discipline to acquire an intellectually firmer foundation. Systems theory, with its technical and seemingly sophisticated vocabulary of "modeling", "mathematics" and "science", seemed to provide this. It was therefore not surprising that it was taken up enthusiastically by a generation of younger planners riding high on the optimism of the 1960s.

Notes

[1] 本文选自：Taylor, N. *Urban Planning Theory since 1945*[M]. Thousand Oaks: Sage, 1998. 尼格尔·泰勒. 1945 年后西方城市规划理论的流变 [M]. 李白玉，陈贞，译. 北京：中国建筑工业出版

社，2006。

[2] 本文主要相关文献：Alexander, 1965; Bruton, 1974; Buchanan, et al, 1963; Chadwick, 1971; Etzioni, 1967; Faludi, 1973a, 1973b; Geddes, 1915; Jacobs, 1961; Keeble, 1952; Kuhn, 1962; Lichfield et al, 1975; Lindblom, 1959; McLoughlin, 1969; Popper, 1963。

[3] 课文中下划线部分为有关城市规划理论的重要论述，需强化阅读，每篇课文习题部分均要求将下划线句子或段落译成中文。

[4] 课文中下划线斜体字部分为复杂长句，其解析详见下文“**Complex Sentences**”。

New Words

accumulate v. 堆积，积累〖前缀 3. ac–; 词根 51. cumul〗
allege v. 断言，宣称
analogous adj. 相似的，可比拟的
arbitrary adj. 随意的，主观的，武断的
assemblage n. 集合，集会〖词根 184. semble〗
attributable adj. 可归因于……的，由……引起的〖词根 214. tribut〗
avow v. 公开宣称，坦率承认
broad-brush adj. 粗枝大叶的，粗线条的
capture v. 俘获；夺取或赢得〖词根 26. cap〗
circulation n. 流通，循环〖前缀 15. circu–〗
coherence n. 连贯性，一致性〖前缀 16. co–; 词根 90. here〗
commitment n. 承诺，保证
compose v. 组成；创作；使平静〖前缀 16. com–; 词根 170. pos〗
comprise v. 包含，包括，由……组成
conceive v. 怀孕；想象，设想
concentrate v. 集中，聚集，专心于〖前缀 16. con–; 词根 32. centr〗
conform v. 一致，遵守〖前缀 16. con–; 词根 76. form〗
cross-fertilization 异花授粉，异化受精
cybernetics n. 控制论
damn v. 诅咒，指责
deficiency n. 缺陷，不足〖前缀 18. de–; 词根 65. fic〗
derive v. 得到，导出；源于，来自
designate v. 指明，指出〖前缀 18. de–; 词根 190. sign〗
dictum n. 名言，格言〖词根 55. dict〗
discrete adj. 个别的，不连续的〖前缀 20. dis–〗

distinguishable adj. 可区别的，可辨别的

dynamism n. 活力，动态

ecological adj. 生态的，生态学的〖词根 57. eco〗

ecosystem n. 生态系统〖词根 57. eco〗

emerge v. 出现，浮现

envisage v. 想象，设想〖前缀 24. en-; 词根 229. vis〗

evolve v. 发展，进化〖词根 234. volv〗

externally adv. 在外部，在外面，外表上〖前缀 26. ex-〗

flexible adj. 灵活的，易弯曲的〖词根 72. flex〗

fluid adj. 流体的，能流动的；不固定的，不稳定的〖词根 75. flu〗

highlight v. 强调，突出

hypothetical adj. 假设的，假定的

illustrate v. 说明，表明〖词根 117. lustr〗

import v. 进口，输入，引进〖前缀 33. im-〗

impressive adj. 给人深刻印象的，感人的〖前缀 33. im-; 词根 172. press〗

inflexible adj. 僵硬的，不可弯曲的〖前缀 34. in-; 词根 72. flex〗

integrate v. 合并，成为一体，融入群体

interaction n. 相互作用，相互影响〖前缀 35. inter-; 词根 2. act〗

interconnected adj. 相互连接的〖前缀 35. inter-〗

internal adj. 内部的〖前缀 34. in-〗

irrelevant adj. 不相关的，不中肯的；离题的，不重要的〖前缀 36. ir-〗

irreparable adj. 不可弥补的，无可挽救的，不能修复的〖前缀 36. ir-〗

literally adv. 逐字地，照字面地；确实地，真正地〖词根 110. liter〗

marginal adj. 边缘的

mark off 划分出

naïve adj. 天真的，幼稚的，轻信的

none the less 仍然，依然

novel adj. 新颖的，新奇的〖词根 141. nov〗

numerous adj. 很多的，许多的〖词根 142. numer〗

once-and-for-all 一次了结地，一劳永逸地，彻底地

original adj. 最初的，原始的；有创意的〖词根 148. ori〗

originate v. 发源，发起；创始，发明〖词根 148. ori〗

purport v. 意指，表明；声称，自称

remit n. 提交；提交审议事项；委托研究的课题

stricture n. 苛评，责难

substantially adv. 本质上，实质上

subsystem n. 子系统，亚系统〖前缀 65. sub-〗

trenchant adj. 一针见血的

triangle n. 三角形〖前缀 72. tri-〗

unity n. 统一，完整〖前缀 76. uni-〗

vivid adj. 生动的〖词根 231. viv〗

water-mark n. 水位标志

Complex Sentences

[1] **One was the "systems view" of planning** noted above, ***which*** *was essentially derived from a theory of the object* ***that*** *town planning seeks to plan, namely, the environment (towns, cities, regions, etc.), now seen as a system of interconnected parts.*

黑体字部分为句子的主干。斜体字部分为 which 引导的定语从句，修饰前面的 the "systems view" of planning，其中，that 引导的定语从句修饰其前的 the object, namely 引导的短语为 the object 的同位语，seen 引导的过去分词短语作定语，修饰前面的 the object。

[2] However, apart from his strictures on the need to undertake surveys prior to planning (see below), **Geddes's ideas remained marginal to the mainstream of town planning thought** throughout the first half of the twentieth century, **which** continued to be dominated by architectural ideas.

黑体字部分为句子的主干，apart from 与 throughout 引导的介词短语作状语，which 引导的从句为定语从句，修饰前面的 the mainstream of town planning thought。

[3] **The impressively sounding language of systems theory and analysis and rigorous statistical methods of investigation** promised ***to place geography on a much firmer theoretical and scientific foundation***, thereby improving its standing as an academic discipline.

正体黑体字部分为主语，斜体黑体字部分为宾语，promised 为谓语，improving 引导的现在分词短语作状语。

[4] So *although* ***the emergence of the systems view of planning*** *in the 1960s* ***can be explained*** *in part as a rational response to the alleged deficiencies of the traditional design-based theory of planning,* **this is only part of the story.**

正体黑体字部分为主句，分句为 although 引导的让步状语从句（斜体字部分），分句主语为 the emergence of the systems view of planning，谓语为 can be explained as，其余部分为宾语。

Exercises

[1] 将课文中下划线句子或段落译成中文。

[2] 读熟或背诵上述复杂长句。

[3] 解析下列单词 [解析单词是指根据单词中包含的词根、前缀推导出单词的含义，例如，apathy（n. 漠然，冷淡），解析式为：a-（无）+ path（感情）+ y →无感情→冷漠，冷淡]。解析的单词选自上文"**New Words**"。

capture v. 俘获；夺取或赢得	irrelevant adj. 不相关的，离题的，不重要的
deficiency n. 缺陷，不足	irreparable adj. 不可弥补的，不能修复的
designate v. 指明，指出	literally adv. 逐字地，照字面地；确实地
illustrate v. 说明，表明	triangle n. 三角形
import v. 进口，输入，引进	unity n. 统一，完整
inflexible adj. 僵硬的，不可弯曲的	vivid adj. 生动的

Unit 10

The Systems and Rational Process Views of Planning (2)

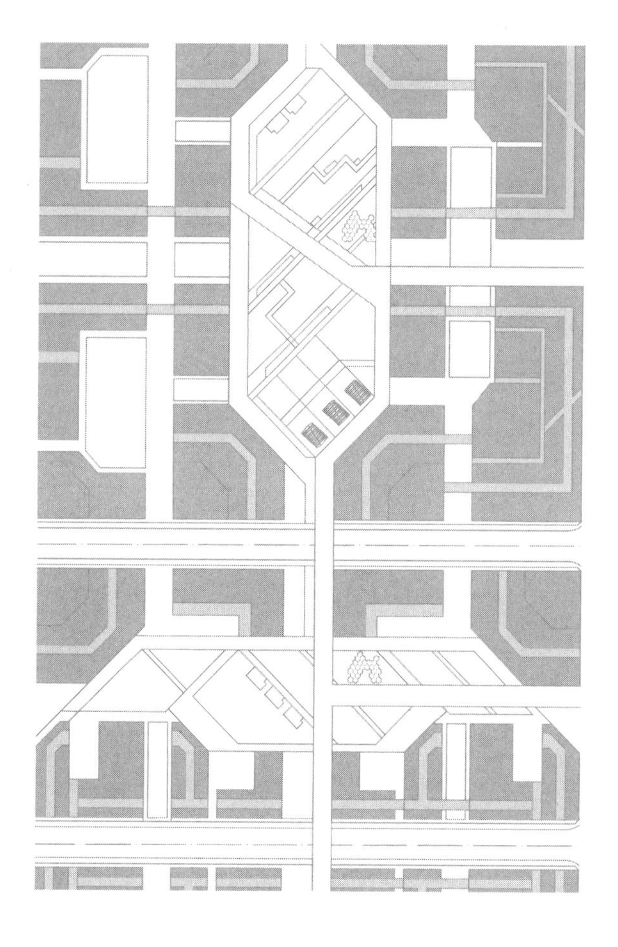

3 The rational process view of planning

3.1 Introduction

Although the systems view of planning derived from a concept of the environment as a system, what this view of planning did not address was the best method, or process, of doing planning. One of the first theorists to draw a distinction between the process of planning and the object or **substance** planning deals with was the American, Melvin Webber: "I understand planning to be a method for reaching decisions, not a body of specific **substantive** goals...planning is a rather special way of deciding which specific goals are to be **pursued** and which specific actions are to be taken...the method is largely independent of the phenomena to be planned" (Webber, 1963) .

This distinction was also emphasized by Andreas Faludi in his book, *Planning Theory* (1973). Faludi similarly distinguishes between "substantive" planning theories about the object (i.e. planning deals with the environment) and "procedural" planning theories about the process or procedures of going about planning. Faludi also described substantive theories as theories in planning, and procedural theories as theories of planning. And since the latter were **literally** theories of planning, Faludi held that planning theory was, or should be, about procedural theory. *Although this idea later attracted criticism (see Chapter 6), what is important to note here is the way the substantive / procedural distinction emphasizes that the systems view of planning, based on a theory of the object (the "substance") planners plan is, in Faludi's terms, a substantive theory, whereas the rational* *process view is clearly a procedural theory.*

It is worth recording that this was not the first theory about the process of planning which had been developed. The Geddesian **dictum** of "survey–analysis–plan" has already been mentioned as a **precursor** to 1960s planning theory, and an analysis of this method may help to explain why the rational process view came to be preferred. The **deficiencies** of survey–analysis–plan (SAP) are threefold. First, to undertake a survey **presupposes** that there is some reason, or purpose, for carrying out the survey. In planning, the suggestion that one needs to undertake a survey to prepare a plan presupposes that there must be some problem one is trying to solve, and that it is the purpose of the survey to **illuminate** this. Logically, therefore, there is a planning "stage" prior to carrying out the survey—namely, the definition of a problem (or an aim). For example, if we undertake a survey of traffic **congestion** prior to preparing a plan, this presupposes that traffic congestion is a problem; and the assumption that traffic congestion is a problem is made logically prior to any survey of it, and it is only in terms of this prior assumption that it "makes sense" to survey traffic congestion.

The second deficiency of SAP is that it implies, by the use of the word "plan" in the singular, that there is only one possible plan, there may be alternative possible strategies. If there are, these alternatives need to be properly evaluated against each. To continue our example of traffic congestion, one plan for solving this problem might involve building new roads and / or widening existing ones. An alternative might involve retaining the existing network of roads and planning to relieve the congestion on them by encouraging the use of bicycles or public transport, etc. If, then, there are possible alternatives like this, it should be part of any rationally considered process of planning to set them out clearly, and then to evaluate them in terms of the problems they are designed to address. Obvious though this is, it is not made clear by the simple formula of "survey–analysis–plan".

Thirdly, in ending with the production of a plan SAP implies that the process of planning ends here. This, too, is **misleading** for we usually seek to implement the plans we make. In other words, a plan is usually followed by some action. "Action" or "implementation" is a further stage of the process of planning not mentioned by SAP. Furthermore, once implemented a plan or policy may turn out to be ineffective or it may have **undesirable** effects which we have not foreseen, etc. So it is also important to monitor the outcomes of our actions to check they are having the effects we want them to have and, if not, we may need to review and revise our actions or plans. To take our example of traffic congestion again, a plan to deal with this problem by building new roads may, when **enacted**, turn out to be fruitless because the new roads may attract more people to travel by car and so lead to a build up of congestion

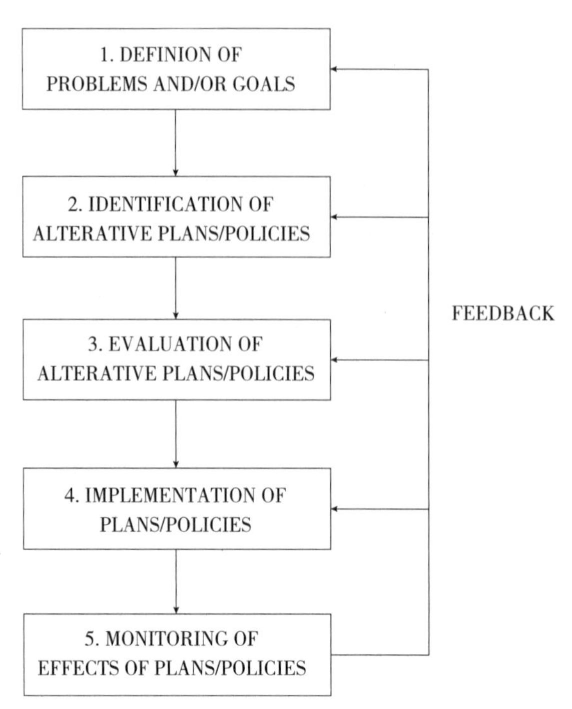

Figure 2 Planning as a process of rational action

again. It is therefore **crucial** to "monitor" the plans we adopt to see whether they are actually solving the problems we want to solve, but again, SAP does not acknowledge this.

From these criticisms we can see what a more considered account of a rational process of planning would have to include. If we **visualize** this process as involving (like the process of "survey–analysis–plan") a number of distinct stages, then we arrive at something like the process shown in Figure 2. Figure 2 distinguishes five main "stages" in a rational process of planning. First there must be some problem or goal which prompts the need for a plan of action. From an analysis of this, a definition of the problems or goals is arrived at. This analysis is necessary not only to guide any empirical investigation (or "surveys") but also because, on closer **inspection**, the initial perceptions of the problems and / or goals may be questionable. It may be that the problem is not really a problem at all, or that what is a problem for one group may not be for another group, or that there are additional problems that were not noticed at first and so on.

The second stage is to consider whether there are alternative ways of solving the problem (or achieving the aim) and, if there are, to clarify these. The third stage is to evaluate which of the **feasible** alternatives is most likely to achieve the desired end. In everyday life we are continually undertaking such evaluations when considering what best to do, and we generally do this "**intuitively**". But in complex decision–making situations the task of evaluating alternatives is, obviously, correspondingly complex, and may require a more systematic analysis of the likely consequences of implementing any alternative. Sophisticated techniques, such as cost–benefit analysis, have therefore been developed for evaluating alternative options, and there is a large literature relevant to this stage of the planning process alone (see, e.g. Lichfield, Kettle and Whitbread, 1975).

The process of planning does not end when a decision has been made, for the chosen policy or plan then needs to be implemented. It is thus more accurate to describe the rational process of planning as a theory or model of rational action, rather than "decision–making". That is why Figure 2 shows "implementation" as a further (fourth) stage in this process. There is yet a fifth stage which involves monitoring the effects of the plan to see whether it achieves the desired ends. A rational process of planning is thus an ongoing or continuous one. Rarely are our objectives achieved perfectly, and even if they are other objectives (or problems) **invariably** arise. Hence the **feedback loops** in Figure 2, indicating that a rational process of planning has no final end–state. Note how the feedback loops may return to any stage of the process: we may need to review our actions, revise our view of the problems, consider other alternatives which we didn't consider before, or accept our initial definition of the options but now question our original evaluation of them and so on.

The recognition that rational planning involved an ongoing, continuous process represented a significant break with the traditional design–based view of town planning. *As noted in Chapter 1, town planners who saw their practice as essentially an exercise in physical design, tended also to see their task as one of producing blueprint plans for towns.* In his book *Planning Theory*, Faludi (1973b) drew a distinction between "blueprint" and "process" planning. The emergence, in the 1960s, of the view of town planning as an ongoing, continuous process (as well as a rational one) therefore implied a rejection of blueprint planning.

3.2 Sources of the rational process view

As with the systems view of planning, the rational process view originated from more general theory which had developed outside the field of town planning. In this case, the relevant theory was "decision theory" (particularly general theory about rational decision–making) which was adopted and applied to town planning (see, e.g. Simon, 1945, 1960; Faludi, 1987).

Two other factors provided a favorable context for the development of the rational process view. In government generally there was an increasing interest in corporate styles of management and decision–making. At the national level this was reflected in the social democratic style of politics which was characterized as a managerial politics in which technical professionals played a key role in advising politicians as to how best to manage the economy, the welfare state, etc. The 1960s was the high tide of the social democratic, "corporatist" state, and the acceptance of rational planning was so widespread that some social theorists, such as Daniel Bell (1960), spoke of the "end of ideology". At local government level, this political stance was reflected in the adoption of strategic and corporate styles of management, following the recommendations of the Bains report (Bains, 1972) into the structure and management of local authorities.

Secondly, there was a renewed faith in the application of "science" to policy making–not only in applying the findings of scientific research to policy but also in relation to the policy-making process itself (hence the talk of the "policy sciences", "scientific management", etc.). Although what made something "scientific" was often not questioned, it was commonly assumed that the quantification of factors relevant to policy (such as traffic flows) was the **hallmark** of "being scientific"; hence, if something could not be quantified (such as the beauty of a place) then it was not considered to be scientific (and hence often **marginalized** in policy-making).

More specifically, **analogies** were drawn between the scientific method as described by the philosopher Karl Popper (e.g. 1963) and the rational process of planning. Popper's account of scientific method insisted that any scientific inquiry was first of all driven by some belief or hypothesis about the world. The role of empirical investigation was then to test the **veracity** of the belief or hypothesis by—according to Popper—examining critically the belief or hypothesis (i.e. attempting to falsify it), so that only those hypotheses which **withstood** critical testing **survived** as **credible** theories. The rational process view of planning was sometimes likened to this. Definitions of planning problems or goals, or even plans and policies themselves, could thus be **equated** with scientific hypotheses that needed to be subjected to severe empirical (or "scientific") testing before being implemented.

3.3 Further clarification of the rational process model of planning

On page 61, I described the basic model of rational decision-making and action which was adopted in the 1960s by those who took a rational process view of planning. **Straightforward** and **uncontroversial** though this model might seem, the idea of rational planning generated considerable debate amongst planning theorists. Some of these were critical of the rational process view, and some of their criticisms are examined in Chapters 5 and 6. Other planning theorists have sought to refine and develop the theory of rational planning, and some of this work is described below as a way of **clarifying** further the rational process view of planning.

3.4 Questions about rationality itself

In the simplest terms, a rational decision (or action) can be described as one for which persuasive reasons can be given. "Persuasive" because "reasons" of some kind can be given for doing almost anything, but having "any" reason for doing something does not **of itself** make that action rational. A persuasive reason is one which connects directly with certain values we hold and aims we wish to attain. In so far as factual evidence is relevant to these values or aims, reasons are also persuasive if they are based on valid knowledge.

However, what **constitutes** a persuasive reason will often be contested. Sometimes there are clusters of persuasive reasons, some of which point in different directions, and so further judgements have to be made about the "balance" of reasons. Moreover, as noted above, if some reasons are persuasive only in terms of certain values, individuals or groups **adhering** to different values may find different reasons persuasive. So adopting something like a rational process approach to planning cannot give us simple, **formulaic** answers to complex problems.

However, the rational process model of planning does imply an approach which gives reasons–and, it is hoped, persuasive ones—for the plans, policies and decisions made. There are three important conditions a rational planning exercise should fulfill. First, the reasons for making planning decisions should be carefully thought through—decisions should be arrived at by considered reflection rather than by guesswork, "**hunch**" or intuition alone. This in turn implies, secondly, that the reasons for making planning decisions should be explicit. If someone were to ask a rational planner why a certain decision had been made, the planner should be able to explain the reasons which led to that decision. Thirdly, if the whole process of planning is rational then each and every stage of a planning process should be carefully and explicitly thought through. The problems or objectives a plan is trying to solve or achieve (i.e. the first stage of the planning process in Figure 2) should be carefully considered and explicitly stated, and likewise the alternative strategies and so on with all the other stages.

3.5 The rational process model as a normative model of planning

So far only a description of the rational planning model has been given; from this it does not necessarily follow that we ought to make decisions in this carefully considered and **explicit** way. Thus some people hold that some kinds of decisions are best made on the basis of intuition–for example, the decisions to marry someone, or have children, or follow a particular career–are all clearly very significant decisions yet some might argue that these are best made on the basis of intuition and "feeling". However, where decisions are being made which significantly affect the lives of others and where there is some form of collective action to implement such decisions, there are strong reasons for insisting that decision-making and action should involve both careful and explicit **deliberation** before policies and plans are adopted (e.g. on the grounds of the numbers of people likely to be affected, the fact that these effects may last a lifetime, etc.). Because town planning is a form of social action, the model of rational planning represents more than just a description of what it might be like to plan rationally; it also represents a model of how we should, ideally, go about planning. The rational process model, therefore, suggests itself as a **normative** model or theory of planning and this was assumed by the rational process theorists of the 1960s and 1970s. As Faludi (1973) put it: "It is only as a normative model that the rational planning process has any meaning at all."

3.6 The rational process model as a model of instrumental (means-end) reasoning

As noted above, the rational process view of planning is clearly a process or procedural theory of planning, not a view or theory about the object or "**substance**" of planning (in Faludi's terms, it is a procedural and not a **substantive** planning theory). Hence the rational process model describes only the "form" (the "procedure") of the reasoning involved in making rational decisions; it says nothing about the actual "substantive" ends or goals planning should aim at. This model simply tells us that, to be rational, a process of decision-making must identify some problem or objective to be solved or aimed at, some plan designed to solve the problems or achieve the objectives and so on. It tells us nothing in itself of which problems to solve, which plans to make, etc. This will be determined by the particular situation. The rational process view of planning is therefore about the method or "means" of planning, not about the "ends" of town planning. This is a "means-end" or "instrumental" model of reasoning, not a model of substantive moral reasoning. Because it was "merely formal", and said nothing about what ends planning should seek to achieve, it was later to come under criticism (see Chapter 6).

3.7 The debate about "disjointed incremental" versus "rational comprehensive" planning

Finally, there has been debate about whether the rational process view of planning necessarily implies the adoption of a "**comprehensive**" approach to planning and policy-making. At first sight, it might seem that rationality does require comprehensiveness, for in order to make perfectly rational decisions, one needs to consider all possible alternatives. *However, limited time in which a decision has to be made, a lack of resources to examine all possibilities, etc., mean that, in reality, it is often impossible to be thoroughly* *comprehensive.* Decision-makers and planners may also simply lack the capacity to absorb and make sense of all the relevant information; they may end up becoming more confused and thus less rational—problems raised by Charles Lindblom (Lindblom, 1959; Braybrooke and Lindblom, 1963). Lindblom proposed an approach which, he claimed, was more relevant to the real world of planning and policy-making. *He suggested that, in most situations, planning has to be* ***piecemeal, incremental, opportunistic*** *and* ***pragmatic****, and that planners who did not or could not operate in these ways were generally* ***ineffective****.* In short, Lindblom presented a model of the "real world" planning as necessarily "**disjointed**" and "incremental", not "rational" and "comprehensive".

It might be thought that, as Faludi (1973) put it, "disjointed incrementalism is highly descriptive of real-life planning", but this does not necessarily mean that this is how planning ought to be approached. Lindblom's critique, however, is not so easily **disposed** of. If real-life planning can only be disjointed and incremental, and since "ought implies can" (i.e. a view about how we ought to act is only persuasive if we can act in this way), Lindblom's suggestion was that a disjointed incremental approach to planning was the only possible approach to planning. It was the best approach we could hope for—much better, in fact, than the impossible ideal of rational comprehensiveness.

The "rational comprehensive" versus "disjointed incremental" debate about planning is considered in Chapter 7. **Suffice** it to say here that the strength of Lindblom's critique rests in our acceptance that being rational requires being comprehensive. If a stranger entered my room with a gun in his hand then I would decide to act quickly. I would therefore have to assess quickly the alternatives open to me and then act. In so doing I would not be **appraising** my options "comprehensively". And yet, given my knowledge of what madmen with guns can do, my decision to **forego** comprehensiveness would be perfectly rational behavior. However, we often cannot afford the **luxury** of examining every possibility because we haven't the time or resources, or because we can only absorb limited information. On such occasions it is perfectly rational to examine, and act upon, the few alternatives we initially perceive as best we can. **Adherents** to the rational model would say that "as best we can" means going through something like the process of decision-making described earlier, however imperfectly. From this point of view, rationality does not require comprehensiveness: in certain circumstances it can be rational to "go through" a rational process of decision-making quickly, even "disjointedly" and "incrementally".

Faludi's (1973) suggestion that rational comprehensive and disjointed incremental planning are alternative modes of planning therefore **poses** a false **dichotomy**. Faludi himself gives various strategies for making rational decisions in conditions when it is impossible to acquire or absorb all relevant information. One was Amitai Etzioni's (1967) idea of "mixed **scanning**", which distinguishes more fundamental or "strategic" decisions from more detailed decisions, and then **advocates** concentrating the process of rational decision-making on the more fundamental decisions. The mixed-scanning approach also involves "**tracking**" the detailed consequences of **crucial**, strategic decisions. In this way the capacity to **oscillate**—or "scan"—between more general or strategic and more detailed levels of thinking is developed.

4 Conclusion: rationality, cybernetics and modernity

This chapter has described two different theories of planning which came to the **fore** in the 1960s. If we accept Faludi's (1973) distinction between **substantive** and procedural planning theories, then the systems view of planning, being a theory about the "**substance**" (the environment) which town planning deals with, was a substantive theory, whilst the rational process view was clearly a procedural theory of planning.

That these two theories are distinct is shown by the fact that it would be possible to **subscribe** to one and not the other. Even if one had not heard of the systems view of planning, one might still believe that planning should be approached in a rational way, and so adopt the rational process model of planning described in this chapter. Indeed, one could imagine this rational process view being combined with the post-war physical design view of planning described in Chapter 1. The fact that the systems and rational process views **emerged simultaneously**, and were often put forward jointly in the new planning textbooks of the 1960s, was thus something of a **coincidence**, but not entirely so. For although they are conceptually distinct, these two theories shared some common **assumptions** which were prevalent in the 1960s.

In this chapter I described some of the general **contextual** factors which help explain why both these theories were taken up by planning theorists in the 1960s. Perhaps the most important of these was the general interest in science and its application to planning and policy-making generally. *One way of describing the change in planning thought which occurred from the 1950s to the 1960s is that, whereas in the 1950s and before, town planning was seen as primarily an art, by the end of the 1960s it had come to be seen as primarily a science.* Both the systems and the rational process views fed off each other in bringing about this change in planning thought. But this interest in science and its practical application had deeper roots.

The two theories of planning described in this chapter can also be viewed as sharing certain fundamental assumptions about the nature of the world and the possibilities for human progress within it. To begin with, both theories were based on a belief in the benefits of planning, and in this they assumed what Giddens (1994) has termed a "cybernetic" model of control. These **presuppositions** were part of a more general set of assumptions which have come to be associated with "modernism"—not just the modern movement in architecture and the arts, but rather a certain way of thinking about the world and social action which had developed in the European Enlightenment of the 18th century. Central to this was a belief in people's capacity to improve the quality of human life on the basis of a scientific understanding of the world. Throughout most of human history, people's lives had been governed by forces which seemed beyond human control and which could strike anyone at any time. The growth of scientific understanding opened up the **prospect** of humanity gaining greater control over the forces of nature and using them to human advantage—in medicine, in agriculture, or in fashioning the environment (it is no accident that landscape design **flourished** in 18th century Europe).

With this there naturally arose a greater optimism about the future—a belief that human life need not be inevitably subject to the **whims** of "fate", but could be improved through rational understanding and action. Twentieth century modernist Utopias, such as Le Corbusier's "radiant city", were expressions of this optimism, as was the more commonplace belief in "progress". So too was the belief that, with a proper scientific understanding of the environment as a "system", coupled with the application of a rational method of decision-making and action, cities and the environment generally could be planned to improve—even "maximize"—human well-being. In this, both the systems and the rational process views of planning were part of the **heady** "modernist" optimism of the 1960s. Indeed, the systems and rational process views of planning can be regarded as marking the high tide of modernist thought—the **crest** of that wave of optimism about the use of science and reason for human progress which had formed the European Enlightenment of the 18th century.

Notes

[1] 本文选自：Taylor, N. *Urban Planning Theory since 1945*[M]. Thousand Oaks: Sage, 1998. 尼格尔·泰勒 . 1945 年后西方城市规划理论的流变 [M]. 李白玉，陈贞，译 . 北京：中国建筑工业出版社，2006。

[2] 本文主要相关文献：Alexander, 1965; Bruton, 1974; Buchanan, et al, 1963; Chadwick, 1971; Etzioni, 1967; Faludi, 1973a, 1973b; Geddes, 1915; Jacobs, 1961; Keeble, 1952; Kuhn, 1962; Lichfield, et al, 1975; Lindblom, 1959; McLoughlin, 1969; Popper, 1963。

[3] 课文中下划线部分为有关城市规划理论的重要论述，需强化阅读，每篇课文习题部分均要求将下划线句子或段落译成中文。

[4] 课文中下划线斜体字部分为复杂长句，其解析详见下文“**Complex Sentences**”。

New Words

adherent n. 支持者，拥护者〖前缀 3. ad–; 词根 90. here〗

adhere v. 黏附；坚持〖前缀 3. ad–; 90. here〗

advocate v. 拥护，支持，鼓吹；n. 支持者，拥护者〖前缀 3. ad–; 词根 232. voc〗

analogy n. 类似，相似，比拟

appraise v. 评价，估计

assumption n. 假定，假设

clarify v. 使清楚，澄清

coincidence n. 巧合，一致〖前缀 16. co–; 词根 23. cid〗

comprehensive adj. 广泛的，综合的

congestion n. 拥挤，堵塞

constitute v. 构成，组成；制定，设立

contextual adj. 文脉上的，前后关系的

credible adj. 可信的，可靠的〖词根 46. cred〗

crest n. 顶点，顶峰

crucial adj. 关键性的，决定性的，十字形的

deficiency n. 缺陷，不足〖前缀 18. de–; 词根 65. fic〗

deliberation n. 考虑，深思熟虑〖前缀 18. de–; 词根 108. liber〗

dichotomy n. 一分为二，对分〖前缀 19. di–〗

dictum n. 名言，格言〖词根 55. dict〗

disjointed adj. 脱离开的，不连贯的〖前缀 20. dis–〗

dispose v. 处理，处置，安排〖前缀 20. dis–; 词根 170. pos〗

emerge v. 出现，浮现

enact v. 制定法律，规定，颁布〖前缀 24. en–; 词根 2. act〗

equate v. 相等，等同〖58. equ〗

explicit adj. 清楚明确的；成熟的，形成的〖前缀 26. ex–; 词根 169. pli〗

feasible adj. 可行的

feedback n. 反馈

flourish v. 繁荣，茂盛〖词根 74. flour〗

fore n. 前部，船头，(马等的) 前腿〖前缀 28. fore–〗

forego v. 放弃，摒弃

formulaic adj. 公式的，刻板的〖词根 76. form〗

hallmark n. 标志，印记

heady adj. 易使人醉的，令人兴奋的

hunch n. 直觉，预感

illuminate v. 照亮，照明；阐明，说明〖词根 116. lumin〗

incremental adj. 增加的〖前缀 34. in−; 词根 45. cre〗

ineffective adj. 无效果的，低效率的

inspection v. 检查，检验，视察，检阅〖前缀 34. in−; 词根 198. spec〗

intuitively adv. 直觉地，直观地

invariable adj. 恒定的，不变的〖前缀 34. in−; 词根 221. vari〗

literally adv. 逐字地，照字面地；确实地，真正地〖词根 110. liter〗

loop n. 圈，环

luxury n. 奢侈，豪华

marginalize v. 使处于边缘，忽视

mislead v. 把……带错路，把……引入歧途〖前缀 41. mis−〗

normative adj. 标准的，规范的〖词根 138. norm〗

of itself 自行

opportunistic adj. 机会主义的，投机取巧的

oscillate v. 摆动，动摇，犹豫

piecemeal adj. 一件一件的，逐渐的；adv. 一件一件地，逐渐地

pose v. 摆好姿势；提出；造成，形成〖词根 170. pos〗

pragmatic adj. 实际的，实用主义的

precursor n. 前辈，前驱，先锋，前任；预兆，先兆〖前缀 57. pre−; 词根 52. cur〗

presuppose v. 假定，假设

presupposition n. 假定，假设

prospect n. 前景，期望〖前缀 58. pro−; 词根 198. spec〗

pursue v. 继续，追求

scan v. 扫描；细看，细查

simultaneously 同时地〖词根 191. simul〗

straightforward adj. 直截了当的，坦率的

subscribe v. 签名，签字；同意，赞成〖前缀 65. sub−; 词根 181. scribe〗

substance n. 物质，材料；实质，本质

substantive adj. 实质的，实体的

suffice v. 足够，满足〖前缀 suf−(=69. sur−); 词根 65. fic〗

survive v. 幸存〖前缀 69. sur–; 词根 231. viv〗

track n. 小路，小道；v. 跟踪，监测

uncontroversial adj. 无争议的，不会引起不和的〖前缀 74. un–; 前缀 17. contro–; 词根 226. vers〗

undesirable adj. 不合意的，不受欢迎的，令人不快的〖前缀 74. un–〗

veracity n. 诚实，真实〖词根 223. ver〗

visualize v. 设想，使可见〖词根 229. vis〗

whim n. 异想天开的念头，怪念头；突然产生的念头，一时的兴致

withstand v. 经受，承受〖前缀 78. with–〗

Complex Sentences

[1] Although this idea later attracted criticism (see Chapter 6), **what is important to note here is the way** *the substantive / procedural distinction emphasizes* ***that the systems view of planning****, based on a theory of the object (the "substance") planners plan* ***is****, in Faludi's terms,* ***a substantive theory, whereas*** *the rational process view is clearly a procedural theory.*

本局结构复杂。正体黑体字部分为句子的主干，以 what 引导的从句作主语。斜体字部分为定语从句，修饰其前的 the way；该定语从句包括一个结构复杂的以 that 引导的宾语从句：宾语从句又分主句和从句，斜体黑体字部分为主句的主干（based 引导的过去分词短语作定语，修饰前面的 the systems view of planning, planners plan 为定语从句，修饰其前的 the object），从句由 whereas 引导，whereas 为连词，表示对比。

[2] As noted in Chapter 1, **town planners** who saw their practice as essentially an exercise in physical design, **tended also to see their task as one of producing blueprint plans for towns**.

本句结构较为简单。黑体字部分为句子的主干，who 引导的从句为定语从句，修饰前面的 town planners。

[3] However, **limited time** in which a decision has to be made, **a lack of resources** to examine all possibilities, **etc., mean that, in reality, it is often impossible to be thoroughly comprehensive**.

黑体字部分为句子的主干（主语、谓语和宾语从句），which 引导的从句为定语从句，修饰前面的 limited time，动词不定式短语作定语，修饰其前的 a lack of resources。

[4] **He suggested that**, in most situations, planning has to be piecemeal, incremental, opportunistic and pragmatic, and **that** planners who did not or could not operate in these ways were generally ineffective.

本句包括两个 that 引导的宾语从句，第二个宾语从句又包括一个 who 引导的定语从句，修饰其前的 planners。

[5] **One way of describing the change in planning thought** which occurred from the 1950s to the

1960s **is** that, whereas in the 1950s and before, town planning was seen as primarily an art, by the end of the 1960s it had come to be seen as primarily a science.

which 引导的从句为定语从句，修饰前面的句子的主语 One way of describing the change in planning thought，that 引导的从句为表语从句，whereas 为连词，表示对比。

Exercises

[1] 将课文中下划线句子或段落译成中文。

[2] 读熟或背诵上述复杂长句。

[3] 解析下列单词 [解析单词是指根据单词中包含的词根、前缀推导出单词的含义，例如，apathy（n. 漠然，冷淡），解析式为：a-（无）+ path（感情）+ y →无感情→冷漠，冷淡]。解析的单词选自上文“**New Words**”。

coincidence n. 巧合，一致	literally adv. 逐字地，照字面地；确实地，
credible adj. 可信的，可靠的	mislead v. 把……带错路，把……引入歧途
dichotomy n. 一分为二，对分	precursor n. 前辈，前驱；预兆，先兆
dictum n. 名言，格言	prospect n. 前景，期望
dispose v. 处理，处置，安排	simultaneously 同时地
enact v. 制定法律，规定，颁布	subscribe v. 签名，签字；同意，赞成
equate v. 相等，等同	suffice v. 足够，满足
explicit adj. 清楚明确的；成熟的，形成的	survive v. 幸存
flourish v. 繁荣，茂盛	veracity n. 诚实，真实

Unit 11

New Planning Paradigms (1)

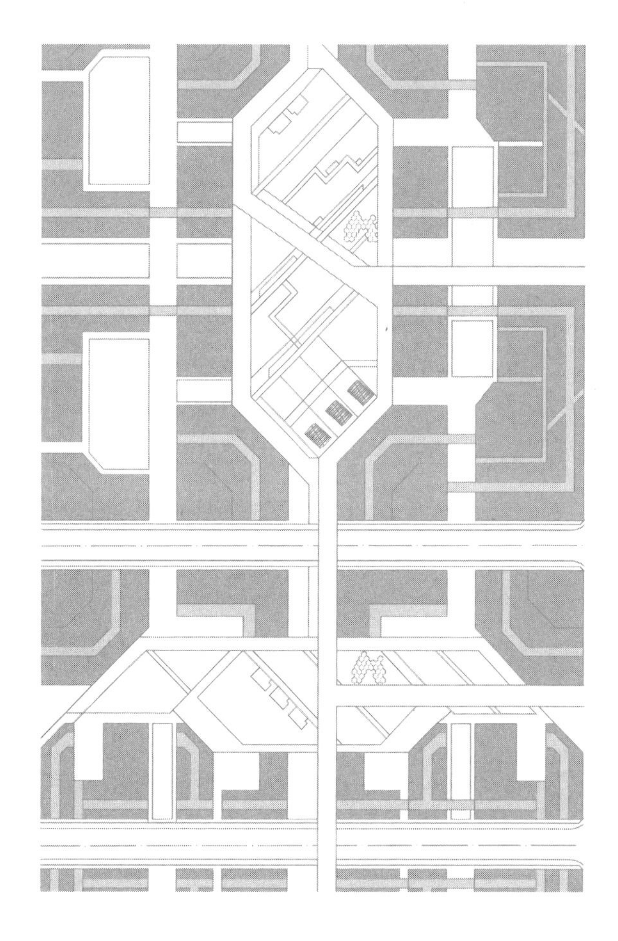

Up to now this book has been an introduction to the problems and the content of spatial planning, treated historically. That has been its aim, as the preface indicated. But now, this last chapter tries to make a bridge to the actual process of planning, as it is carried out by progressive planning authorities at the present time. This process is based strongly on theoretical concepts, which are well set out in modern textbooks of planning. Therefore, this chapter, which tries to **distil** the central content of these more advanced texts, will perhaps serve as an introduction to them for the student of planning.

We need in this to **distinguish** three quite separate stages in the **evolution** of planning theory. *The first, developed from the earliest times down to the mid–1960s–and well exemplified in the early development plans coming after "the 1947 Town and Country Planning Act"–could be called the master plan or **blueprint** era.* The second was **ushered in** from about 1960, and **replaced** the first approach through "the Planning Advisory Group (PAG) of 1965" and "the 1968 Town and Country Planning Act"; it could be called the systems view of planning. The third, which began to **evolve** in the late 1960s and the 1970s, is more **heterogeneous** and more **diffuse**; it may best be **labeled** the idea of planning as continuous **participation** in **conflict**. In what follows I shall first describe the **transition** from blueprint to systems planning, and then the more complex transition to **participative**–conflict planning.

1 Systems planning versus master planning

The change that occurred after 1960 was based on the notion that all sorts of planning

constitute a distinct type of human activity, concerned with controlling particular systems. Thus spatial planning (or, as it is called here, urban and regional planning) is just a subclass of a general activity called planning; it is concerned with managing and controlling a particular system, the urban and regional system. *It follows from this that all planning is a continuous process which works by seeking to devise* ***appropriate*** *ways of controlling the system concerned, and then by* ***monitoring*** *the effects to see how far the controls have been* ***effective*** *or how far they need* ***subsequent modification***. This view of planning is quite different from the one held by an older generation of planners, such as Geddes or Abercrombie, or even the generation which set up the planning system in Britain after the Second World War. These older planners saw planning as concerned with the production of plans which gave a detailed picture of some desired future end state to be achieved in a certain number of years. It is true that under "the 1947 Planning Act" in Britain, **deliberate provision** was made for review of the plans every five years. But the philosophy behind the process was heavily **oriented** towards the concept of the fixed master plan.

Arising from this basic difference of approach, there were also detailed differences between 1940s and 1960s planning. The old planning was concerned to set out the desired future end state in detail, in terms of land–use patterns on the ground; the new approach, **embodied** in Britain in the new structure plans prepared under "the 1968 Planning Act", **concentrates** instead on the **objectives** of the plan and on **alternative** ways of reaching them, all set out in writing rather than in detailed maps. Again, the old planning tended to **proceed** through a simple sequence **derived** from Patrick Geddes: survey–analysis–plan. The existing situation would be surveyed; analysis of the survey would show the **remedial** actions that needed to be taken; the fixed plan would embody these actions. But in the new planning the emphasis is on tracing the possible consequences of alternative policies, only then evaluating them against the objectives in order to choose a preferred course of action; and, it should be emphasized, this process will continually be repeated as the monitoring process throws up **divergences** between the planner's intentions and the actual state of the system.

The new concept of planning derived from one of the newest sciences: **cybernetics**, which was first identified and named in 1948 by the great American mathematician and thinker Norbert Wiener. Rather than dealing with a completely new subject matter, cybernetics is essentially a new way of organizing existing knowledge about a very wide range of phenomena. Its central notion is that many such phenomena—whether they are social, economic, biological or physical in character—can usefully be viewed as complex **interacting** systems. The behavior of atomic particles, a jet aeroplane, a nation's economy—

all can be viewed, and described, in terms of systems; their different parts can be separated, and the interactions between them can be analyzed. Then, by introducing appropriate control mechanisms, the behavior of the system can be **altered** in specific ways, to achieve certain objectives on the part of the controller. The point here is that it is necessary to understand the operation of the system as a whole (though not necessarily in complete detail throughout) in order to control it **effectively**; unless this is done, actions taken to control one part of the system may have completely unexpected effects elsewhere. A good example is the design of a motor car; if the designer produces **extra** power without considering the total impact on the rest of the complex system that makes up the car, the result could be **instability** or rapid **wear** of other parts, with disastrous results.

Cybernetics has already had considerable practical applications in modern technology, especially in the complex control systems which monitor spacecraft or **automatic** power stations. Its applications to the world of social and economic life are still **tentative**. Some observers think that human mass behavior is too complex and too **unpredictable** to be reduced to cybernetic laws. Others find **ethically repellent** the idea that planners should seek to control the operation of people as if they were machines. All that can be said with certainty is that in some areas where people and machines interact—as for instance in urban traffic control systems—cybernetics is already proving its **effectiveness**. It still remains to be proved definitively whether the application can be **extended equally** well to all areas of human behavior.

Fundamental to the concept of systems planning—as the cybernetics-based planning has come to be called—is the idea of interaction between two parallel systems: the planning or controlling system itself, and the system (or systems) which it seeks to control. This notion of constant interaction should be kept in mind throughout the following account of the systematic planning process. More particularly, we are concerned with this process as it applies to spatial planning, using the word "spatial" in its widest sense: it need not be limited to the three-dimensional space of **Euclidean geometry**, but may extend for instance, to include notions of economic space (the costs involved in **traversing** distance) and **psychological** or perception space. Nevertheless, there can be little doubt that in some sense, however **distorted** by psychological or economic factors, the relationship of parts of the urban and regional system in geographical space must be the central concern of the urban and regional planner.

To control these relationships, in a mixed economy such as the United States or the countries of Western Europe, the planner has two main **levers**: one is the power to control public investment, especially in elements of **infrastructure** such as roads, railways, airports, schools, hospitals and public housing schemes; the other is the power to encourage or discourage **initiatives** from the private sector for physical development, through **incentives** or **disincentives** to industrial development, controls on land use, and environmental regulations. Both these forms of power, of course, vary in their scope and **effectiveness** from one nation or society to another. Different countries invest different proportions of their gross national product in public infrastructure (though in advanced industrial countries there are limits to this variation); different nations have very widely differing controls over physical development (though in none, apparently, is there either a complete lack of such controls, or a completely effective central control). Therefore, almost by definition, the urban or regional planner will never be completely **ineffective**, or completely **omnipotent**. The planner will exist in a state of continuous interaction with the system s/he is planning, a system which changes partly, but not entirely, as a result of processes beyond the planner's mechanisms of control.

Against this background, it is now possible to appreciate the ***schematic summaries*** *of the planning process set out by three leading British* ***exponents*** *of the systematic planning approach: Brian McLoughlin, George Chadwick and Alan Wilson.* McLoughlin's account (Figure 3a) is the simplest; it proceeds in a straight line through a **sequence** of processes, which are then constantly **reiterated** through a return **loop**. Having taken a basic decision to adopt planning and to set up a particular system, planners then formulate broad goals and identify more detailed objectives which logically follow from these goals. They then try to follow the consequences of possible courses of action which they might take, with the aid of models which simplify the operation of the system. Then they evaluate the alternatives in relation to their objectives and the resources available. Finally, they take action (through public investment or controls on private investment, as already described) to **implement** the **preferred** alternative. After an **interval** they review the state of the system to see how far it is **departing** from the assumed course, and on the basis of this review they begin to go through the process again.

Chadwick's account of the process is **essentially** a more complex account of the same sequence (Figure 3b). Here, a clear distinction is made between the observation of the system under control (the right–hand side of the diagram) and the planners' actions in devising and testing their control measures (the left–hand side). **Appropriately**, there are return loops on both sides of the diagram, indicating again that the whole process is **cyclical**. But at each stage of the process, in addition, the planners have to **interrelate** their observations of the system with the development of the control measures they intend to apply to it.

Wilson's account (Figure 3c) is even more theoretically complex, but again it can be related to Chadwick's. In it there are not two sides of the process which interact, but three levels presented vertically. The most basic level, corresponding to part of Chadwick's right–

hand sequence, is simply called "understanding" (or, in the **terminology** of the American planner Britton Harris, "prediction"). It is concerned wholly with devising the working tools, in the form of techniques and models, which are needed for the analysis of the system under control. The **intermediate** level, corresponding to another part of Chadwick's right–hand side, is concerned with the further use of these techniques in analyzing problems and **synthesizing** alternatives which will be **internally** consistent. The upper level, corresponding roughly to the left–hand side of the Chadwick diagram, is essentially concerned with the positive actions which the planner takes to regulate or control the system: goal formulation, evaluation of

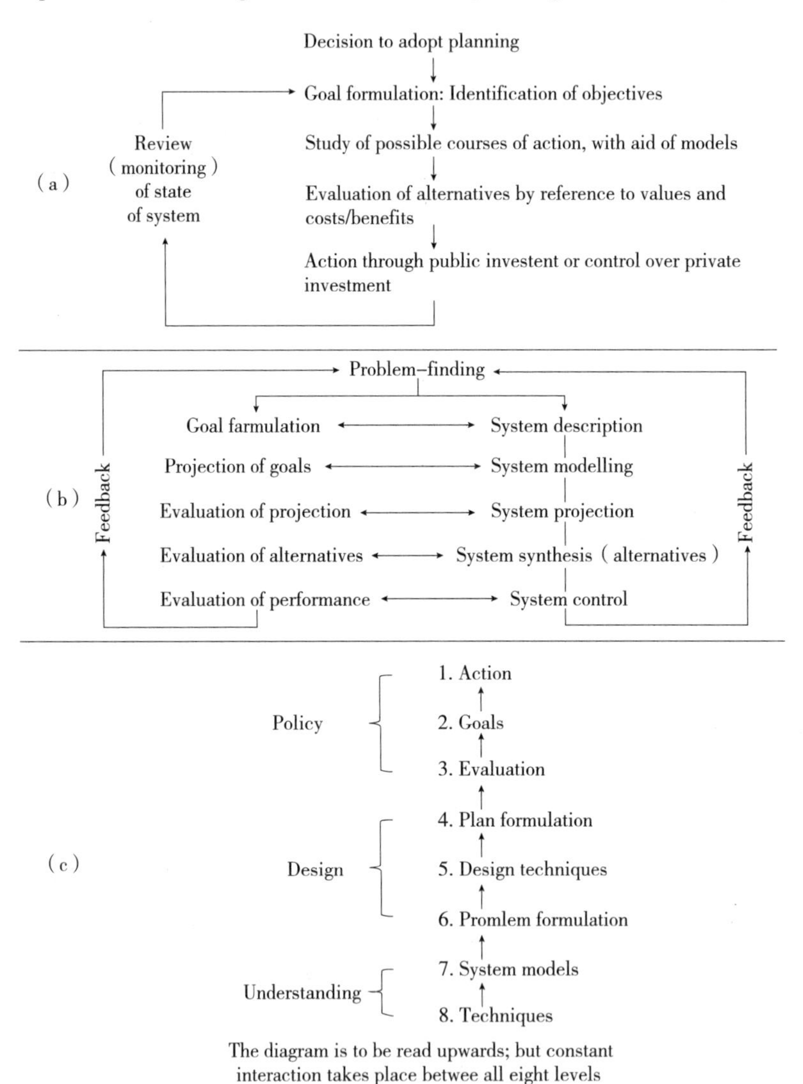

Figure 3 Three concepts of the planning process: (a) Brian Mcloughlin; (b) George Chadwick; (c) Alan Wilson. During 1960s interest developed in systematizing the process of planning, with a new stress on modeling and evaluating alternative designs or courses of action. These formulations drew heavily from the science of cybernetics and system analysis.

alternatives, and actual implementation of the preferred alternative.

All three accounts are helpful ways of looking at the planning process. But since simplicity must be the **essence** of this **summary** chapter, the following accounts of the separate stages of the process are based principally on the classification of Brian McLoughlin.

Notes

[1] 本文选自：Hall，P. Urban and Regional Planning [M]. 4th edition. London and New York: Routledge，2002. 彼得・霍尔．城市和区域规划（原著第四版）[M]. 邹德慈，李浩，陈熳莎，译．北京：中国建筑工业出版社，2008。

[2] 主要相关文献：Castells, 1977; Chadwick, 1971; Faludi, 1973; Forester, 1993, 1999; Friedmann, 1988; Hall, 1988; Harvey, 1973; Healey, 1988, 1997; McLoughlin, 1969。

[3] 课文中下划线部分为有关城市规划理论的重要论述，需强化阅读，每篇课文习题部分均要求将下划线句子或段落译成中文。

[4] 课文中下划线斜体字部分为复杂长句，其解析详见下文“**Complex Sentences**”。

New Words

alter v. 改变，更改〖词根 7. alter〗

alternative adj. 择一的，供选择的；n. 选择，替换物〖词根 7. alter〗

appropriate adj. 适当的，恰当的

appropriately adv. 适当地

automatic a. 自动的，无意识的；n. 小型自动武器〖前缀 9. auto–〗

blueprint n. 蓝图，设计图

concentrate v. 集中，聚集，专心于〖前缀 16. con–; 词根 32. centr〗

conflict 冲突，战斗〖前缀 16. con–; 73. flict〗

constitute v. 构成，组成；制定，设立

cybernetics n. 控制论

cyclical adj. 循环的

deliberate adj. 故意的，深思熟虑的〖前缀 18. de–; 词根 108. liber〗

depart v. 离开，起程，出发

derive v. 得到，导出；源于，来自

diffuse v. 漫射，发散；adj. 散开的〖前缀 dif–(=20. dis–); 词根 80. fus〗

disincentive adj. 阻止的，抑制的；n. 起抑制作用的事物，制止因素

distil v. 蒸馏，提炼

distinguish v. 区分，辨别

distort v. 扭曲，弄歪〖词根 211. tort〗

divergence n. 分叉，分歧，离题〖前缀 19. di–; 词根 225. verg〗

effective adj. 有效的，起作用的

effectively adv. 有效地

effectiveness n. 有效，有效性

embody v. 表现，体现〖前缀 23. em–〗

equally adv. 平等地，公正地〖词根 58. equ〗

essence n. 本质，精髓

essentially adv. 本质上，基本上

ethical adj. 道德的

Euclidean adj. 欧几里得的，欧几里得几何学的

evolution n. 发展，进化〖词根 234. volu〗

exponent n. 倡导者，拥护者

extend v. 延伸，扩大，推广〖前缀 26. ex–; 词根 208. tend〗

extra adj. 额外的，特别的；adv. 特别地；n. 额外的事物，额外费用〖前缀 27. extra–〗

geometry n. 几何学〖词根 geo, 地；词根 126. metr〗

heterogeneous adj. 异类的，不同的〖前缀 29. hetero–; 词根 81. gene〗

implement v. 实施，执行

incentive n. 动机，刺激，诱因

ineffective adj. 无效果的，低效率的

infrastructure n. 基础设施，基础建设〖词根 201. struct〗

initiative n. 倡议，主动性，主动权

instability n. 不稳定，不稳固〖前缀 34. in–〗

interaction n. 相互作用，相互影响〖前缀 35. inter–; 词根 2. act〗

intermediate adj. 中间的，中级的〖前缀 35. inter–; 词根 122. medi〗

internal adj. 内的，内部的〖前缀 34. in–〗

interrelate v. 相互关联，相互影响〖前缀 35. inter–〗

interval n. 间隔

label n. 标签，标记，符号；v. 贴标签于，把……称为，把……列为

lever n. 杠杆；方法，手段〖词根 106. lev〗

loop n. 圈，环

modification n. 修改，修正

monitor v. 监控，监听，监视

objective adj. 客观的，真实的；目标的，目的的；n. 目标，目的

omnipotent adj. 全能的，万能的〖前缀 46. omni-; 词根 171. pot〗

orient v. 标定方向，使……向东方；adj. 东方的，新生的；n. 东方〖词根 148. ori〗

participation n. 参与，参加

participative adj. 参与的，下级职员参与决策的

prefer v. 更喜欢

proceed v. 进行，前进〖前缀 58. pro-; 词根 30. ceed〗

provision n. 规定，条款；供应

psychological adj. 心理学的，心理的，精神上的

reiterate v. 重申；反复地做

remedial adj. 治疗的，医治的

repellent adj. 令人厌恶的〖前缀 61. re-; 词根 157. pel〗

replace v. 把……放回原位；替换，代替〖前缀 61. re-〗

schematic adj. 纲要的；计划的

sequence n. 序列，顺序，连续〖词根 187. sequ〗

subsequent adj. 随后的，跟随的〖前缀 65. sub-; 词根 187. sequ〗

summary n. 摘要，概要；adj. 概括的，总结的〖词根 202. sum〗

synthesize v. 合成〖前缀 70. syn-〗

tentative adj. 试验的，尝试性的

terminology n. 专门名词，术语

transition n. 过渡，转变，变迁〖前缀 71. trans-〗

traverse v. 横贯，横越；通过

unpredictable adj. 无法预言的，不可预测的〖前缀 74. un-; 前缀 57. pre-; 词根 55. dict〗

usher v. 预报……的到来；宣告，显示 (in)

wear n. 磨损

Complex Sentences

[1] **The first**, developed from the earliest times down to the mid-1960s—and well exemplified in the early development plans coming after "the 1947 Town and Country Planning Act"-**could be called the master plan or blueprint era**.

黑体字部分为句子的主干，developed 引导的过去分词短语作定语，修饰其前的 The first，两个破折号之间的部分为解释性插入语。

[2] **It follows from this *that*** *all planning is a continuous process* ***which*** *works by seeking to devise appropriate ways of controlling the system concerned, and then by monitoring the effects to see how far the controls have been effective or how far they need subsequent modification.*

斜体字部分为 that 引导的主语从句（It 为形式主语），其中 which 引导的从句为定语从句，修饰其前的 a continuous process。

[3] Against this background, **it is now possible to appreciate the schematic summaries of the planning process** set out by three leading British exponents of the systematic planning approach: Brian McLoughlin, George Chadwick and Alan Wilson.

黑体字部分为句子的主干，set out 引导的过去分词短语作定语，修饰前面的 the schematic summaries of the planning process，冒号后为解释性插入语。

Exercises

[1] 将课文中下划线句子或段落译成中文。

[2] 读熟或背诵上述复杂长句。

[3] 解析下列单词［解析单词是指根据单词中包含的词根、前缀推导出单词的含义，例如，apathy（n. 漠然，冷淡），解析式为：a-（无）+ path（感情）+ y →无感情→冷漠，冷淡］。解析的单词选自上文"**New Words**"。

conflict 冲突，战斗	omnipotent adj. 全能的，万能的
diffuse v. 漫射，发散；adj. 散开的	proceed v. 进行，前进
distort v. 扭曲，弄歪	repellent adj. 令人厌恶的
divergence n. 分叉，分歧，离题	sequence n. 序列，顺序，连续
heterogeneous adj. 异类的，不同的	subsequent adj. 随后的，跟随的
interaction n. 相互作用，相互影响	transition n. 过渡，转变，变迁
lever n. 杠杆；方法，手段	unpredictable adj. 无法预言的，不可预测的

Unit 12

New Planning Paradigms (2)

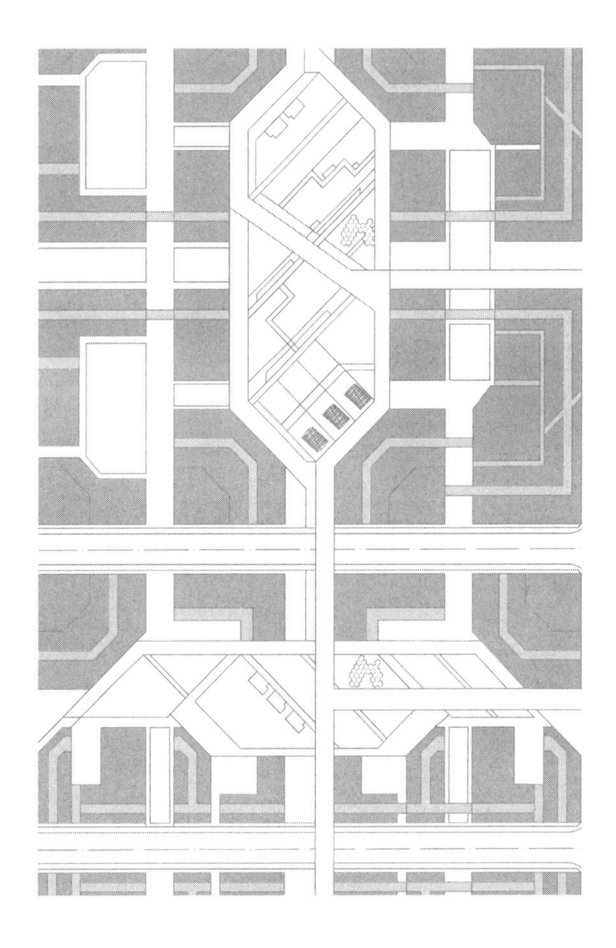

2 Goals, objectives and targets

Planning, as a general activity, may have one objective or many. There is no necessary relationship between the scale and expense of a planning program, and the complexity of the objectives behind it; thus the American moon–shot program, one of the costliest pieces of investment in the history of humankind, had a fairly obvious single main objective. Most urban and regional planning activities, however, have **multiple** objectives. *The first step in the planning process, then, is to identify those purposes which the planner seeks to achieve, to order them in terms of their importance, and to consider how far they are* ***reconcilable*** *each with the other.* This might seem obvious, yet surprisingly, most plans of the past prove to be very **perfunctory** in their treatment of objectives; it seems almost as if the aims of the plan were so well understood that no one needed to set them down. But unless objectives are made **explicit**, no one can be sure that they are shared by the people they are being planned for; nor is it possible **rationally** to prefer one plan to another.

Modern plan **methodology**, therefore, lays great stress on this first step in the process. In particular, it distinguishes rather carefully between three stages in the development of aims: goal formulation, identification of objectives, and target–setting. Goals are essentially general and highly abstract; they tend to fall into broad categories such as social, economic and aesthetic (some of which categories may overlap), and they may include qualities of the planning process itself, such as **flexibility**. Some authors, notably Wilson, define goals in a rather different way, as areas of concern; in this view, planners start by identifying broad functional sub-systems which are of interest to them, because they appear to present problems which may be **amenable** to the controls they propose to manage. Examples of areas of concern would include public health, education, income and its distribution, **mobility** (both physical and social), and environmental quality. Objectives in contrast are rather more specific; they are defined in terms of actual programs capable of being carried into action, though they fall short of detailed quantification. They also require the expenditure of resources (using that word in its widest sense, to include not merely conventional economic resources but also elements like information) so that they **imply** an element of competition for scarce resources. Thus if "mobility" is a general goal, the resulting objectives might include a reduction of travel time in the journey to work, an improvement in the quality of public transport (or of a part of it), or a program of motorway construction to keep pace with rising car ownership. Notably, as in the cases just quoted, objectives can be devised only as the result of a more detailed **scanning** of the system being planned, in order to identify specific **malfunctioning** or **deficiencies**. Finally, as a further stage of refinement, objectives are turned into **targets** representing specific programs in which criteria of performance are set against target dates. *Thus the detailed targets developed from the above objectives might include construction of a new underground railway line within ten years to reduce journey times in the north-western sector of the city by an average of 20 per cent; or construction of a new motorway link within five years in order to cut traffic delays by some specific amount.* Targets, by their nature, tend to be very specific and particular; one problem that emerges from the whole goals-objectives-targets process, therefore, is that of **integrating** rather **disparate** individual programs into a **coherent** plan.

Already, this first stage in the planning process **involves** great difficulties of a conceptual and technical nature. In the first place, it is not entirely clear who should take the lead in the process. Broad goals for society, it might be argued, are a matter for the politicians, though the professional planner can play a valuable role by trying to order the choices. But politicians are largely involved with **acute** short-term issues; their timescale is very different from that of the planner, whose decisions may have an impact for generations. The public themselves form a very **heterogeneous** mass of different groups whose value systems are almost certainly very different if they are not in open **contradiction**. Even the identification of these groups **poses** difficulties, because most people will belong to more than one group for different purposes; they will have interests and values as members of families living at home, as workers in a factory or office, as consumers, and perhaps as members of voluntary organizations, and the values of these groups may actually come into **conflict** with each other. Public opinion polls and other surveys may throw limited and **distorted** light on preferences, because most people find difficulty in thinking about highly abstract goals that do not concern them immediately, and because they will not easily imagine long-term possibilities outside their immediate range of experience. Because of differences and even conflicts of view, it is almost certainly impossible even to devise a satisfactory general welfare function which would somehow combine all the individual preferences and weightings of different individuals or groups.

It is no wonder, then, that in his comments on goal formulation Chadwick points out that "the gap between theory and possible practice is pretty wide". Planners do the best they can by trying to **amass** as much information as possible about their **clients** and their values; by trying to identify acknowledged problem areas, where by fairly common agreement something needs to be done; and by using logical argument to proceed from general goals to more specific objectives. **Evolving** research tools, such as simulation and gaming—whereby members of the public are faced with imaginary choice situations which test their preferences—will also help to throw light on one particular dark area: the weighting of different objectives and the **trade-off** between them. But it should not be expected that there will be a dramatic breakthrough in this intellectually very difficult area.

3 Forecasting and modeling

Having defined objectives and given them some precise form in the shape of targets based on performance criteria, planners will turn to description and analysis of the urban or regional system they wish to control. *Their aim here is to find ways of representing the behavior of the system over time—both in the recent past and in the future—in such a way that they can understand the impact of alternative courses of action that are open to them.* To do this, they will produce a model of the system (or, more likely, a number of **interconnected** models which seek to describe the behavior of its subsystems). A model is simply a **schematic** but precise description of the system, which appears to fit its past behavior and which can, therefore, be used, it is hoped, to predict the future. It may be very simple: a statement that population is growing by 2 per cent a year is in effect a model of population growth. But it may be, and often is, **computationally** quite complex.

There are two important questions that the planner needs to **resolve** about the modeling process: first, what aspects of the urban system s/he wishes to model; second, what sorts of model are available. The answer to the first question will, of course, depend on the planner's precise interests; the planner must first say what questions the model is required to answer. But usually, the urban and regional planner is concerned with the spatial behavior of the economy or of society. In particular, s/he is interested in the relationships between social and economic activities—such as working, living, shopping, and enjoying recreation—and the spaces (or structures) available to house them. The planner will need to know the size and location of both, as well as the **interrelationships** between activities (transportation and communication) which use special spaces called channel spaces (roads or railways, telephone wires). Together, these aspects of the urban system can be said to **constitute** activity systems. Particularly important among them, for the urban planner, is the relationship between workplaces, homes, shops and other services, and the transportation system that links these three.

The answer to the second question—the choice of type of model—will again depend on the object of the planning exercise. Models, whether simple or complex, are capable of being classified in a number of different ways. They may be **deterministic** in character, or **probabilistic** (i.e. **incorporating** an element of chance). They may be **static** in character, or **dynamic**. Many of the best-known urban development models are static; that is, they project the system only for one future point in time, at which point the system is regarded as somehow reaching **equilibrium**. This, of course, is a totally unrealistic assumption which is not supported by knowledge of how the system actually behaves, and one of the main challenges is to produce better dynamic models which are usable. Another separate but related question is whether the model chosen is to be simply descriptive of the present (or recent past) situation, or predictive of the future, or even **prescriptive** in the sense that it contains some element of built-in evaluation. Self-evaluating models are not very common in urban and regional planning, though they do exist: the linear programming model, which **automatically maximizes** the achievement of some variable subject to certain **constraints**, is the most notable example and has been used in planning contexts both in the United States and in Israel. But more commonly the model merely predicts the future; it can be run a number of times with different policy assumptions underlying it, but finally the choice will be made through a quite separate evaluative process.

Yet another question is the choice between spatially **aggregated** models and spatially **disaggregated** models. A model which projects some sub-system for the town or region as a whole is termed spatially aggregated; a model which examines the **internal** zone-by-zone **allocation** of that system is spatially disaggregated. Urban and regional planners, of course, require both sorts of model, but the results of their spatially disaggregated models must **accord** with the control totals given by the spatially aggregated ones. Well-known population projection models, such as the **cohort survival** model (which operates through the survivorship rates of **successive** five-year age cohorts of the population), are spatially aggregated; so are the common economic models, such as input-output models. Models which predict future distributions of people and service industries within urban areas, such as the well-known Garin–Lowry model used in many planning studies, are, of course, spatially disaggregated.

Model design is one of the most complex and **intriguing** stages of the modern planning process. Designing a model, or models, to **suit** the precise problem **involves** logical analysis of a set of **interrelated** questions. Once it is determined precisely which questions the model is supposed to answer, the problem is to list the concepts to be represented, which must be measurable. It is also necessary to investigate which **variables** can be controlled by the planner, at least in part; if the assumption is that no parts are controllable, then the model is a pure forecasting model, but if at least some of the factors are under the planner's control, then it is a planning model. The planner must also consider what behavioral theories about systems are to be embodied in his or her model. S/he must consider technical questions, questions such as how the variables are to be **categorized** or **subdivided** (as, for instance, population can be categorized by age, sex, occupation or industry group); how **explicitly** time will be treated; and how the model is to be **calibrated** and tested. The answers to these questions will depend in part on the techniques that are available, and on the relevant data that can be used to **illustrate** them, as well as on the computational capacity of the computer which will be used to run the models. Fortunately, with the increasing power of personal computers, this tends to be no longer a **constraint**.

4 Plan design and plan evaluation

Many standard accounts of the modern planning process refer to a stage which is called plan design, or plan formulation. *To the **lay** observer, this would appear to be the critical point where, when all technical aids have been used to the utmost, the planner takes command and exercises his or her creative abilities, just as s/he did in a simpler age before computer modeling had become an **integral** part of the planning process.* In an important way, this is true: there must be at least one point in the whole process, and in all probability more than one point, where the planner exercises a power to **synthesize disparate** elements into a **coherent** plan. But in fact this power has to be **manipulated** in close relationship to the machine. What the computer—and above all the personal computer—has done is to speed up, many times, the power to generate, and to evaluate, alternative formulations of the plan. The capacity to design is essentially the capacity to use this power critically and creatively.

The design process, therefore, really starts as soon as the planner begins to design the models. At that point, the critical questions—what elements of the urban system should the models represent, and in how much detail—will finally determine the content of the plan design. To all intents and purposes, the model is the design, and alternative assumptions built into the model generate alternative design possibilities. Of course, the word "design" here is not being used in a conventional sense. In most cases the urban and regional planner does not end by producing a blueprint for actual physical structures on the ground. What s/he tries to do is to specify a future state or states of the urban and regional system which appear, from the operational model, to be **internally** coherent and consistent, and to be workable and **feasible**; and which also best satisfy the objectives which have been set. The content of the design, and of the model which **embodies** it, will depend on the focus and the objectives of the planner. Thus if the plan stresses transportation, it will chiefly consist of a design for channel spaces to accommodate projected traffic flows. If the plan stresses social **provision**, it will embody locations for social service facilities in relation to the distribution of projected demands from different sections of the population. **Invariably**, following the modern stress on planning as a process, the design will not be a **one-shot** plan for some target date in the future; rather, the model or models which **incorporate** the design will represent a continuous **trajectory** from the present into the predictable future.

Design therefore essentially consists of two elements. The first is the choice of system models to represent the main elements which the design should incorporate, and the running of these models to give a number of coherent and realistic pictures of the future state of the system through time. The second is the process of evaluation of the alternatives to give a **preferred** or **optimum** solution. At the stage of evaluation, the goals and objectives which the planner has generated are applied directly to the alternative simulations of the future system.

Like most other terms in the planning process, the word "evaluation" needs careful definition. To most lay observers it **conveys** a **connotation** of economic criteria: evaluation, crudely, represents the best plan for money. Many notable modern planning exercises have in fact made **extensive** use of economic evaluation procedures; some of these will be described in summary a little later. But essentially, evaluation consists of any process which seeks to order preferences. Strictly speaking, it need not refer to money values, or to use of economic resources, at all.

What is essential is that evaluation **derives** clearly from the goals and objectives set early on in the planning process. The first question must be how well each design alternative meets these objectives, either in a general sense or (preferably) in terms of satisfying quantified performance criteria. Very commonly it is found that many objectives contain an element of **contradiction** in practice. It is difficult, for instance, to **reconcile** the objective "preserve open countryside" with the objective "give people the **maximum** freedom to enjoy the private environment they want", or alternatively to reconcile "provide for free movement for the car-owning public" with "preserve the urban **fabric**". Somewhere along the line, either in the **original** formulation of objectives or in the evaluation process, it is necessary for the planning team to devise weights which rank some objectives above others, and indicate how much different objectives are worth in relation to each other. This may involve a conscious decision to favor one group of the **client** population more than another, because quite often the interests of these groups are in **conflict**: car-owners versus non-car-owners, for instance, or old-established rural residents versus new interests. Such value judgements are hard to make, and the political process must **inevitably** have a large hand in them.

To try to make plan evaluation more **rigorous**, since about 1955 at least three techniques have gained widespread currency in the planning world. The best known of these among the general public, cost-benefit analysis, is **explicitly** economic in its approach. The second best-known evaluative device in planning, Professor Nathaniel Lichfield's Planning Balance Sheet, is essentially a modified cost-benefit analysis. The Goals Achievement Matrix of Professor Morris Hill is the third of the evaluation devices which have gained currency in urban and regional planning.

One important question about the whole plan-design process is whether it should be linear or **cyclical**. The version so far developed in this chapter is linear: that is, the alternative plans are developed and modeled, all in equal detail, up to the point where they are all evaluated side by side with a common set of evaluation procedures. In fact, several major British planning exercises—such as the sub-regional studies for Nottinghamshire—Derbyshire and for Coventry, Solihull and Warwickshire—have instead used a cyclical approach. A number of very crude alternatives are developed, modeled and evaluated. Certain among them are **eliminated**, but one or more are **retained**, and these (or combinations and **permutations** of them) are developed and modeled in greater detail. This process may be repeated three or four times, with the modeling-evaluation process progressively testing finer and more subtle **variations** of detail. The cyclical or **recursive** approach appears more complex, particularly when it is applied in the plan report. But it can be argued that it is more economical of the planning team's skills and of computer time, and by logically eliminating alternatives and concentrating on detailed variations it acts as a systematic educative process for the team.

5 Implementing the plan

By systematic evaluation of alternatives, the planner can select a preferred course of action for **implementation**. But it needs to be stressed again that this is no **once-and-for-all** decision. In the planning process outlined here, the whole exercise of modeling, evaluation and selection is continuously repeated. *The objective is to have on the one hand a* ***monitoring*** *system which checks the response of the urban and regional system to the various planning measures taken to control its progress; and on the other hand the control system itself, which responds* ***flexibly*** *and* ***sensitively*** *to the information controlled by the monitoring system.* The **analogy**, of course, is with piloting a ship or an airplane. A course is set; a battery of instruments **confirms** that the craft is on course, or that it is **deviating** from course; appropriate control devices, either **automatic** or **manual**, take appropriate corrective action. The monitoring system thus tests the **correspondence** (or lack of correspondence) between the real-world situation and the model (or navigation chart) that has been set up to describe it. If there is a **divergence**, then either controls must be operated to bring the real-world situation again in **conformity** with the model design, or the model must be altered to make it a more realistic description of the way the world works, or some combination of the two.

The above is **frankly** a description of a planning ideal rather than of present planning reality anywhere. The world that urban and regional planning seeks to control is much bigger and richer in content than the rather limited piece of reality represented by the course of a ship or an aircraft. To reduce it to schematic terms by means of a model is **correspondingly** more difficult, and the likelihood of error much greater. Because of the complexity of the human resources involved, the control systems open to the planner are much cruder and less effective than those available to the ship's captain or airline pilot. The history **recounted** earlier in this book proves definitely that even in strong and effective planning systems, the world changes in all kinds of ways that planners fail to predict, so that plans may fail **grievously** to correspond to reality, even after a very few years. In any event, even if we did know how to control the responses in the urban and regional system, to exert pressure effectively might prove politically **unfeasible**.

In practice, as revealed in some notable planning **controversies** of recent years, a **tidy** systems view of planning may go wrong in a variety of ways. In the first place, knowledge about the **external** environment of the planning decision may increase rapidly, with **unpredictable** results. The changing economies of nuclear and conventional power production may **invalidate** a program of power-station location; variations in the noise **emission** levels of jet airplanes, and improvements in ground-level city-to-airport transport may completely change the framework of a decision about airport location (while a change in the size of planes may make a new airport unnecessary); the development of quieter, or completely silent, cars might **render** many current controversies about motorways irrelevant. In practice it ought to be possible to predict technical changes and their impact rather better than is now generally done. (It seems extraordinary, for instance, that just after the Second World War, when jet aircraft were already flying, their noise impact seems to have been **ignored** in planning all the major civil airports of the world.) But even so, it must be recognized that there will always be a considerable element of **unpredictability** and chance.

Second, plans can go wrong because of the complex interrelationships between different levels of the planning system, and between different elements in the planning situation. Thus a general, high-level strategic policy may be laid down by a national or regional planning authority for apparently good cause, but with unexpected effects at the more local planning level. In Britain office development policy was a good example: it was introduced in 1964–1965 with the aim of **restricting** office growth in London and other major cities, and of **promoting decentralization** to new towns and development areas. But the restrictions had the effect of holding up for many years some important pieces of redevelopment in London, such as that of Piccadilly Circus, which depended for their commercial **viability** on the office content. The process can, however, work in **reverse**. Thus, though almost everyone was agreed on the need for a national motorway from London to south Wales to relieve traffic **congestion** on the old road, work on the new highway was held up for over ten years as one local **amenity** society after another successfully **diverted** the line of the road from its own area. The fact that as finally built, the motorway probably follows the least environmentally damaging line is perhaps some **consolation**.

Third, there is the fact that over time human values—or at least the values of those actively concerned—tend to change. In recent years there is evidence that the pace of such changes is actually increasing; fashions in planning tend to change almost as fast as fashions in clothing. Since complex plans inevitably take time to prepare and then to execute, the result may be controversy. Urban redevelopment provides a good example. In the late 1950s and early 1960s the key word was "comprehensive redevelopment": to provide a better environment and separate people's activities from the danger and pollution of traffic, it was necessary to make a clean sweep of many old urban areas. But by the late 1960s and early 1970s there had been virtually a **reversal**: because of influence by the highly **persuasive** book by Jane Jacobs, *The Death and Life of Great American Cities* (see the list of further reading at the end of Chapter 8), the key words became "conservation" and "urban **spontaneity**", and younger planners in particular wanted to keep the chaos and disorder of the older city, which they saw as attractive. *Plans which represented the older scheme of values, such as the Greater London Council's redevelopment of Covent Garden* *or the reconstruction of the La Defense area of Paris, were bitterly attacked for just those qualities which would have made them admired a few years before.* Similarly, the late 1960s saw a **revulsion** against motorway-building in cities, with protests as far apart as San Francisco and London, New Orleans and Paris. Earlier, it had been almost **axiomatic** that urban traffic should be channeled on to special segregated routes designed for the purpose. But now, **opponents** began to stress the disadvantages of the motorways: noise, visual **intrusion**, and **severance** of traditional neighborhoods. Since it was impossible for the city ever to cater adequately for the rising tide of car traffic, objectors argued, the right policy was to restrict the use of cars in cities and build up good public transport instead. And by the end of the 1990s, there was an even more remarkable reversal: planners, influenced by the American "new urbanism", began to argue that the entire notion of a **hierarchical** street system was wrong, and that it was best to plan urban streets for more or less equal **permeability** to traffic. But this was modified in turn by an even more profound and persuasive shift: traffic experts now accepted that to build one's way out of traffic congestion was **self-defeating**, since—after some delay—new traffic would simply be generated to fill the available new space, leaving **congestion** much as before.

Finally, however, the problem is that it is very difficult to reconcile different sets of values. Most planning controversies, even though the bitterness of the debate may **obscure** the fact, involve a conflict of right against right. Other things being equal, it would be right to build urban motorways to cater for traffic; if it can be assumed that there is no way of stopping people buying and using cars, and that in fact these cars do provide desirable personal mobility, then urban motorways are the best way of handling the resulting problem. The trouble is that this is not the only consideration. As opponents are not slow to point out, motorways are **intrusive** and **disruptive**, even if better designed than most are (and better design is certainly possible to achieve); funds spent on them may well be **diverted** away from public transport; even when the great majority of households own cars, as has been the case since the mid-1950s in the United States and since the mid-1980s in Britain, the great majority of individuals at most times will still be without free **access** to one; and, as just noted, the benefits will soon be **swamped** by further traffic growth. The controversy, then, is essentially about priorities. In a perfect world without an economic problem, there would be unlimited resources for very well-designed motorways, **integrated** into the urban fabric, and for an equally **superb** public transport system available to all—not to mention all the other competing investments such as the replacement of old schools and mental hospitals and prisons, and the construction of new homes for those who are still **inadequately** housed. But of course the resources are far from unlimited; and the community as a whole has to decide which of many good things it wants the most.

In the final analysis, therefore, most major planning decisions are political in character. Unfortunately, as is well known, political decision-making is a highly **imperfect** art. Ordinary people are given the choice of voting every four or five years for a national government, and perhaps every three years for a local government; in either case, they must vote on a confusing **bundle** of different policies, in which planning issues are well down the list. Many of these issues, as stressed more than once in this chapter, may be so general and abstract in character that it is difficult for the ordinary citizen to appreciate their impact until critical—and perhaps **irrevocable**—decisions have been taken. Pressure groups may achieve effective action on particular issues, but they tend to be formed and populated **disproportionately** by those groups in society that are better educated, better informed and better organized—which, in most cases, also means richer. The recommendations of the Roskill Commission on London's third airport were finally **overruled** by the minister after a great public **outcry**; and many planners thought the minister right. But many also took little comfort in the fact that whereas the commission's work had cost just over £1 million, the pressure group against it spent £750, 000 in getting the recommendation **overturned**. Likewise, the inquiry in the late 1990s into Terminal 5 at London's Heathrow Airport, which took four years and cost £70 million, mainly benefited lawyers and **convinced** the government that a simpler way must be found, almost certainly through debates on national transport issues in Parliament. The danger here is that the greater the call for public participation in planning decisions, the greater the likelihood that decisions will go in favor of the richer and better organized—and against those who can least look after their own interests. **Disillusioned**, in the 1980s and 1990s some environmental activists took to the streets and even to the trees, practising direct action to stop construction of major highways like the M3 Winchester bypass or the A34 Newbury bypass. They did not succeed, but they almost certainly brought about the **abandonment** of other schemes like the East London River Crossing and associated roads. The critical question here was: whose interests did they represent, and in what way did their actions fit into an ordered and rational and democratic decision-making? These questions are far from being satisfactorily answered.

A particularly **acute** problem of **divergent** values, which is evident in many planning decisions, concerns the **trade-off** between the interests of different generations. In such situations the best is **literally** the enemy of the good. Should public housing, for instance, be built to reflect the standards and **aspirations** of the first generation of occupiers, or the second and third? If built merely to minimal **contemporary** standards, the risk is that it will be regarded as sub-standard within a generation or two; and it may not then be possible to redesign it except at unacceptable cost. But if it is built in advance to satisfy the standards of tomorrow, then fewer resources will be available to satisfy the pressing needs of today. Similarly, many decisions about preservation and conservation involve questions of the interests of different generations. It may be cheaper to pull down a Georgian housing area in a city and replace it by new flats than to **rehabilitate** it; the community is then faced with a choice between the needs of those who are ill-housed, and the value of the area for generations of future citizens. Similarly, the establishment of green belts around British cities after the Second World War involved certain sacrifices on the part of those who were thereby housed farther away from their jobs in the cities, while the majority of the urban populations of that time were unable to enjoy any benefits because they lacked the cars to make **excursions** into the protected countryside; the true benefits will probably be experienced by the next generation, who will use cars to travel out to new country parks established in the green belts long after they were **designated**. Here, planners may with justification claim that by their **intervention** they are guarding the interests of **posterity**, including generations yet unborn. But if fundamental values change from one generation to another—if, for instance, each generation values environmental conservation more highly than its **predecessor**—how is that resolved? And suppose for instance that values vary geographically, so that unemployed people in a **depressed** region or town care more for job creation than the environment, while rich people in a more **affluent** city value the environment more—how are those differences to be reflected and **accommodated**?

Planning in practice, however well managed, is therefore a long way from the **tidy sequences** of the theorists. *It involves the basic difficulty, even impossibility, of predicting future events; the interaction of decisions made in different policy spheres;* ***conflicts*** *of values which cannot be fully resolved by rational decision or by calculation; the* ***clash*** *of organized pressure groups and the defence of* ***vested*** *interests; and the* ***inevitable confusions*** *that arise from the complex interrelation-ships between decisions at different levels and at different scales, at different points in time.* The **cybernetic** or systems view of planning is a condition towards which planners aim; it will never become complete reality.

Notes

[1] 本文选自：Hall，P. Urban and Regional Planning[M]. 4th edition. London and New York: Routledge, 2002. 彼得·霍尔 . 城市和区域规划（原著第四版）[M]. 邹德慈，李浩，陈熳莎，译 . 北京：中国建筑工业出版社，2008。

[2] 主要相关文献：Castells, 1977; Chadwick, 1971; Faludi, 1973; Forester, 1993, 1999; Friedmann, 1988;

Hall, 1988; Harvey, 1973; Healey, 1988, 1997; McLoughlin, 1969。

[3] 课文中下划线部分为有关城市规划理论的重要论述，需强化阅读，每篇课文习题部分均要求将下划线句子或段落译成中文。

[4] 课文中下划线斜体字部分为复杂长句，其解析详见下文“**Complex Sentences**”。

New Words

abandonment n. 放弃，放纵〖前缀 1. a–; 词根 18. band〗

accommodate v. 容纳，向……提供；使适应，使和谐一致

accord v. / n. 同意，一致〖前缀 3. ac–; 词根 41. cord〗

acute adj. 敏锐的，灵敏的；(病)急性的〖词根 1. acu〗

affluent adj. 富裕的，富足的〖前缀 3. af–; 词根 75. flu〗

aggregate adj. 聚集的，合计的；n. & v. 总数，合计〖前缀 3. ag–; 词根 86. greg〗

allocation n. 配给，分配

amass v. 积聚

amenable adj. 愿服从的，通情达理的

amenity n. 便利设施

analogy n. 类似，相似，比拟

aspiration n. 强烈的愿望〖词根 200. spir〗

automatic a. 自动的，无意识的；n. 小型自动武器〖前缀 9. auto–〗

automatically adv. 自动地〖前缀 9. auto–〗

axiomatic adj. 公理的，自明的

bundle n. 捆，批

calibrate v. 量……口径；校准

categorize v. 把……分类

clash v. & n. 冲突

client n. 委托人，当事人，顾客

coherent adj. 一致的，连贯的〖前缀 16. co–; 词根 90. here〗

cohort n. 有相同统计要素(如年龄等)的一组人；同伴，同谋者

computational adj. 计算的

confirm v. 证实，证明〖前缀 16. con–; 词根 69. firm〗

conflict n. 冲突，战斗〖前缀 16. con–; 词根 73. flict〗

conformity n. 一致，遵从，顺从〖前缀 16. con–; 词根 76. form〗

confusion n. 混乱，混淆〖前缀 16. con–; 词根 80. fus〗

congestion n. 拥挤，堵塞

connotation n. 言外之意，内涵〖前缀 16. con–; 词根 139. not〗

consolation n. 安慰，慰问〖前缀 16. con–; 词根 194. sol〗

constitute v. 构成，组成；制定，设立

constraint n. 强制，强迫；对感情的压抑

contemporary adj. 当代的，同时代的〖前缀 16. con–; 词根 207. tempor〗

contradiction n. 反驳；矛盾〖前缀 17. contra–; 词根 55. dict〗

controversy n. 公开辩论，论战〖前缀 17. contro–; 词根 226. vers〗

convey v. 运载，运送；表达

convince v. 使某人确信，说服

correspondence n. 信件，通信；相应

corresponding adj. 相应的，符合的

cybernetics n. 控制论

cyclical adj. 循环的

decentralization n. 分散〖前缀 18. de–; 词根 32. centr〗

deficiency n. 缺陷，不足〖前缀 18. de–; 词根 65. fic〗

depressed adj. 情绪低落的，沮丧的；萧条的〖前缀 18. de–; 词根 172. press〗

derive v. 得到，导出；源于，来自

designate v. 指明，指出〖前缀 18. de–; 词根 190. sign〗

deterministic adj. 确定性的

deviate v. 越轨，脱离〖前缀 18. de–; 词根 227. vi〗

disaggregate v. 分解〖前缀 20. dis–; 前缀 3. ag–; 词根 86. greg〗

disillusion v. 梦想破灭，醒悟〖前缀 20. dis–〗

disparate adj. 迥然不同的〖前缀 20. dis–; 词根 151. par〗

disproportionate adj. 不成比例的，不相称的〖前缀 20. dis–〗

disruptive adj. 分裂性的，破坏的；扰乱的〖前缀 20. dis–; 词根 178. rupt〗

distort v. 扭曲，弄歪〖词根 211. tort〗

divergence n. 分叉，分歧，离题〖前缀 19. di–; 词根 225. verg〗

divergent adj. 分叉的，叉开的；发散的，扩散的〖前缀 19. di–; 词根 225. verg〗

divert v. 使转移〖词根 26. vert〗

dynamic n. 动态；动力

eliminate v. 排除，消除

embody v. 表现，体现〖前缀 23. em–〗

emission n. 排放，散发〖前缀 e–(= 26. ex–); 词根 129. miss〗

equilibrium n. 平衡〖词根 58. equ〗

evolve v. 发展，进化〖词根 234. volv〗

excursion n. 短途旅游〖前缀 26. ex–; 词根 52. cur〗

explicit adj. 清楚明确的；成熟的，形成的〖前缀 26. ex–; 词根 169. pli〗

explicitly adv. 明白地，明确地〖前缀 26. ex–; 词根 169. pli〗

extensive adj. 广大的，广阔的；多方面的，广泛的〖前缀 26. ex–; 词根 208. tens〗

external adj. 外面的，外部的〖前缀 26. ex–〗

fabric n. 织物；构造，结构

feasible adj. 可行的

flexibility n. 灵活性，机动性，柔韧性〖词根 72. flex〗

frankly adj. 直率地，坦诚地

grievous adj. 令人悲痛的，使人伤心的

heterogeneous adj. 异类的，不同的〖前缀 29. hetero–; 词根 81. gene〗

hierarchical adj. 等级的，分层的〖词根 14. arch(y)〗

ignore v. 忽视，不顾〖词根 82. gnor〗

illustrate v. 说明，表明〖词根 117. lustr〗

imperfect adj. 有缺点的，不完美的〖前缀 33. im–〗

implementation n. 实施，执行

imply v. 暗示〖前缀 33. im–; 词根 169. pli〗

inadequate adj. 不充足的；不适当的〖前缀 34. in–〗

incorporate v. 合并，并入〖前缀 34. in–; 词根 42. corpor〗

inevitable adj. 不可避免的，必然的〖前缀 34. in–〗

integral adj. 构成整体所必需的，完整的

integrate v. 合并，成为一体，融入群体

interconnected adj. 相互连接的〖前缀 35. inter–〗

internal adj. 内部的〖前缀 34. in–〗

interrelated adj. 相互关联的〖前缀 35. inter–〗

interrelationship n. 相互关系〖前缀 35. inter–〗

intervention n. 干预，干涉，介入〖前缀 35. inter–; 词根 222. vent〗

intrigue v. 密谋；引起极大兴趣

intriguing adj. 引起兴趣或好奇心的，吸引人的

intrusion n. 侵扰，干扰〖前缀 34. in–; 词根 216. trus〗

intrusively adv. 入侵地〖前缀 34. in–; 词根 216. trus〗

invalidate v. 使无效，使作废〖前缀 34. in–〗

invariable adj. 恒定的，不变的〖前缀 34. in–; 词根 221. vari〗

involve v. 包含，牵扯，使参与〖前缀 34. in–; 词根 234. volv〗

irrevocable adj. 无法取消的，不能改变的〖前缀 36. ir–; 前缀 61. re–; 词根

232. voc 〗

lay adj. 世俗的；外行的，没有经验的

literally adv. 逐字地，照字面地；确实地，真正地〖词根 110. liter〗

malfunction v. 发生故障；n. 故障，障碍〖前缀 38. mal-; 词根 79. funct〗

manipulate v. 操作，处理〖词根 119. man〗

manual adj. 手的，手工的；体力的〖词根 119. manu〗

maximize v. 最大化，使增至最大限度

maximum adj. 最大值的，最大量的

methodology n. 方法学，方法论

mobility n. 流动性，移动性〖词根 130. mob〗

monitor v. 监控，监督，监听

multiple adj. 多重的，多个的，复杂的〖前缀 43. multi-〗

once-and-for-all adj. 一劳永逸的

one-shot adj. 一次完成的，一次有效的，一次性的

opponent n. 对手，敌手

optimum adj. 最适宜的

original adj. 最初的，原始的；有创意的〖词根 148. ori〗

outcry n. 大声疾呼，抗议〖前缀 47. out-〗

overrule v. 驳回，否决；统治，支配〖前缀 48. over-〗

overturn v. 颠覆，推翻〖前缀 48. over-〗

perfunctory adj. 敷衍的，草率的〖前缀 53. per-; 词根 79. funct〗

permeability n. 渗透性〖前缀 53. per-〗

permutation v. 改变，改动；互换，交换〖词根 135. mut〗

persuasive adj. 有说服力的，劝说的，劝诱的

posterity n. 后代，子孙〖前缀 56. post-〗

predecessor v. 前辈，前任〖前缀 57. pre-; 词根 30. cess〗

prefer v. 更喜欢

prescriptive adj. 规定的，指定的，命令的，规范的

probabilistic adj. 概率性的；盖然性的，或然性的

promote v. 促进，推进，提升〖前缀 58. pro-; 词根 132. mot〗

provision n. 规定，条款；供应

rationally adv. 理性地

reconcilable adj. 可调和的，可和解的

reconcile v. 调停，使和解

recount v. 详细叙述

recursive adj. 有重复能力的，可重复使用的〖前缀 61. re–; 词根 52. cur〗

rehabilitate v. 修复，恢复

render v. 使成为；递交；给予；表达

resolve v. 决定，决意；解决，解答

restrict v. 限制，约束

retain v. 保持

reversal n. 反向，反转，倒转〖前缀 61. re–; 词根 226. vers〗

reverse adj. 反面的，颠倒的；v. 反转，颠倒〖前缀 61. re–; 词根 226. vers〗

revulsion n. 厌恶，憎恶；剧烈反应〖前缀 61. re–; 词根 175. vuls〗

rigorous adj. 严格的，严密的

scan v. 扫描；细看，细查

schematic adj. 纲要的；计划的

self–defeating adj. 自我挫败的，违背自己利益的，自我拆台的〖前缀 64. self–〗

sensitively adv. 敏感地，慎重地〖词根 186. sens〗

sequence n. 序列，顺序，连续〖词根 187. sequ〗

severance n. 切断，断绝〖前缀 63. se–〗

spontaneity n. 自发，自生

static adj. 静止的，不变的

subdivide v. 再分，细分〖前缀 65. sub–〗

successive adj. 连续的，相继的；接替的，继承的〖前缀 66. suc–; 词根 30. cess〗

suit v. 适合，适宜

superb adj. 上乘的，出色的〖前缀 68. super–〗

survival n. 幸存〖前缀 69 sur–; 词根 231. viv〗

swamp v. 淹没，使沉默；n. 沼泽，沼泽地

synthesize v. 合成〖前缀 70. syn–〗

target n. 目标，目的

tidy adj. 整洁的，整齐的；有条不紊的，严紧的

trade–off n. 协调，平衡；交换

trajectory n. 弹道，轨道

unfeasible adj. 不能实行的，难实施的〖前缀 74. un–〗

unpredictability n. 不可预测性，不可预知性〖前缀 74. un–; 前缀 57. pre–; 词根 55. dict〗

unpredictable adj. 无法预言的，不可预测的〖前缀 74. un–; 前缀 57. pre–; 词根 55. dict〗

variable adj. 变化的，可变的；n. 变量〖词根 221. vari〗

variation n. 变化，变动〖词根 221. vari〗

vest n. 马甲，防护衣；v. 穿衣服；授权，赋予

viability n. 生存能力，生活力〖词根 227. vi〗

Complex Sentences

[1] **The first step in the planning process, then, is** *to identify those purposes which the planner seeks to achieve, to order them in terms of their importance, and to consider how far they are reconcilable each with the other.*

黑体字部分为主语、谓语，斜体字部分为表语，三个表语（三个动词不定式短语），第一表语含有一个 which 引导的定语从句，第三表语含有一个 how far 引导的宾语从句。

[2] Thus **the detailed targets** developed from the above objectives **might include** construction of a new underground railway line within ten years to reduce journey times in the north-western sector of the city by an average of 20 per cent; or construction of a new motorway link within five years in order to cut traffic delays by some specific amount.

黑体字部分为主语、谓语，两个 construction 短语作宾语，即两个宾语，developed 引导的过去分词短语作定语，修饰前面的 the detailed targets。

[3] **Their aim here is to find ways of representing the behavior of the system over time**–both in the recent past and in the future–in such a way **that** they can understand the impact of alternative courses of action **that** are open to them.

黑体字部分为句子的主干，两个破折号之间为解释性插入语，其后的两个 that 引导的从句均为定语从句，第一个定语从句修饰其前的 such a way, 第二个定语从句修饰其前的 alternative courses of action。

[4] To the lay observer, **this would appear to be the critical point** ***where***, ***when*** *all technical aids have been used to the utmost, the planner takes command and exercises his or her creative abilities,* **just as** s/he did in a simpler age **before** computer modeling had become an integral part of the planning process.

黑体字部分为句子的主干。斜体字部分为 where 引导的定语从句，修饰其前的 the critical point，该定语从句又包括一个 when 引导的时间状语从句。just as 引导的从句为比较状语从句，其中又包括一个 before 引导的时间状语从句。

[5] **The objective is to have on the one hand a monitoring system** which checks the response of the urban and regional system to the various planning measures taken to control its progress; **and on the other hand the control system itself**, which responds flexibly and sensitively to the information controlled by the monitoring system.

黑体字部分为句子的主干，注意 on the one hand...on the other hand... 结构；第一个 which 引导的定语从句修饰其前的 a monitoring system，第二个 which 引导的定语从句修饰其前的 the control system。

[6] **Plans** which represented the older scheme of values, such as the Greater London Council's redevelopment of Covent Garden or the reconstruction of the La Defense area of Paris, **were bitterly attacked for just those qualities** which would have made them admired a few years before.

黑体字部分为句子的主干，第一个 which 引导的定语从句修饰其前的 Plans，第二个 which 引导的定语从句修饰其前的 those qualities。

[7] **It involves** *the basic difficulty, even impossibility, of predicting future events*; *the interaction of decisions made in different policy spheres*; *conflicts of values which cannot be fully resolved by rational decision or by calculation*; *the clash of organized pressure groups and the defence of vested interests*; *and the inevitable confusions that arise from the complex interrelationships between decisions at different levels and at different scales, at different points in time.*

黑体字部分为主语和谓语，斜体字部分为宾语，六个宾语：第二宾语包括一个过去分词短语定语，第三宾语包括一个 which 引导的定语从句，第六宾语包括一个 that 引导的定语从句。

Exercises

[1] 将课文中下划线句子或段落译成中文。

[2] 读熟或背诵上述复杂长句。

[3] 解析下列单词 [解析单词是指根据单词中包含的词根、前缀推导出单词的含义，例如，apathy（n. 漠然，冷淡），解析式为：a-（无）+ path（感情）+ y →无感情→冷漠，冷淡]。解析的单词选自上文“**New Words**”。

abandonment n. 放弃，放纵	intervention n. 干预，干涉，介入
accord v. / n. 同意，一致	intrusion n. 侵扰，干扰
affluent adj. 富裕的，富足的	irrevocable adj. 无法取消的，不能改变的
aggregate v. 集合，合计	malfunction v. 发生故障；n. 故障，障碍
consolation n. 安慰，慰问	manipulate v. 操作，处理
deviate v. 越轨，脱离	perfunctory adj. 敷衍的，草率的
disaggregate v. 分解	permeability n. 渗透性
disruptive adj. 分裂性的，破坏的；扰乱的	permutation v. 改变，改动；互换，交换
distort v. 扭曲，弄歪	predecessor v. 前辈，前任
divert v. 使转移	reversal n. 反向，反转，倒转
equilibrium n. 平衡	revulsion n. 厌恶，憎恶；剧烈反应
excursion n. 短途旅游	sequence n. 序列，顺序，连续
explicit adj. 清楚明确的；成熟的，形成的	severance n. 切断，断绝
heterogeneous adj. 异类的，不同的	superb adj. 上乘的，出色的
hierarchical adj. 等级的，分层的	survival n. 幸存

Unit 13

New Planning Paradigms (3)

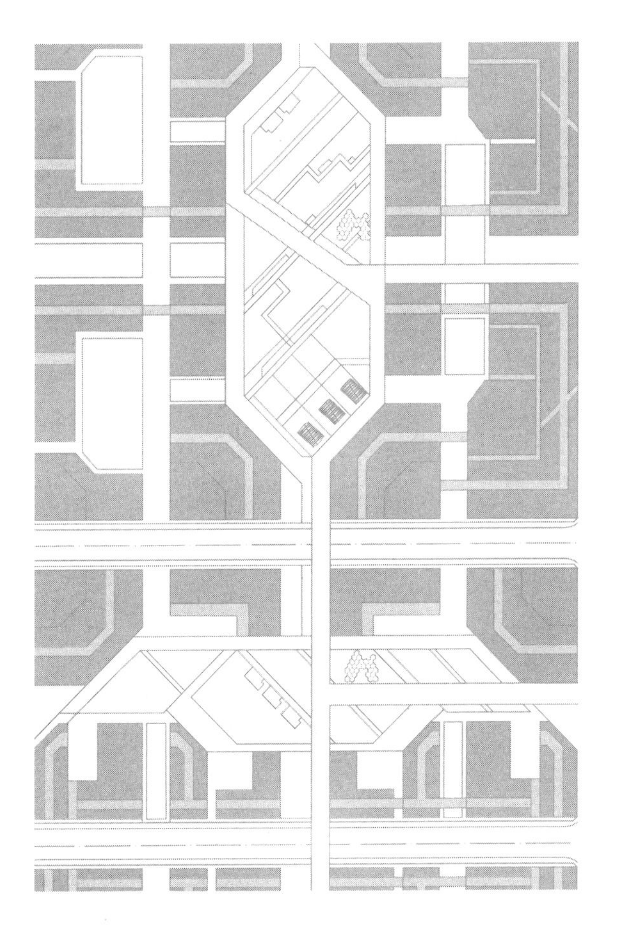

6 New planning paradigms

Because of these difficulties, as was perhaps inevitable, in the first half of the 1970s there was a major **reaction** against the style of systems planning—just as, a few years before, the systems planners had **reacted** against the master planners. *In particular, some planners began to question the basic* **tenets** *of the systems approach: the notion that it was scientific, in the sense that the world could be completely understood and its future states predicted; the notion that planning could be value free, in that the planner could* ***disinterestedly*** *determine what was best for society; the notion that the planner was planning for a society that was a* **homogeneous aggregate**, *in which the welfare of the entire people was to be* ***maximized***, *without too much concern with distributional questions; and the notion that the task of planning was to come to terms with—which, in practice, mean adapting to—the facts of rapid growth and change.* These ideas had proved particularly timely in two kinds of planning which **tackled** major problems of the 1960s: transportation planning, to deal with the facts of explosive car ownership; and sub-regional planning, to deal with the **equally pressing** facts of population growth and **decentralization**. Though it was **subsequently criticized** on technical grounds, there can be no doubt that in these fields—and in the structure plans of the early 1970s—the systems approach represented a considerable advance on the older, **inflexible** style of planning. In the more **stagnant** and **constricted** world of the 1970s, however, its concepts and techniques appeared to lose some of their point.

The problem, though, went deeper than that, and the attack on systems planning came earlier. First, there was the demand for public **participation** in planning. Beginning with official **endorsement** in the Skeffington Report of 1969, which resulted in a **statutory** requirement that participation be formally **incorporated** into the planning process, it struck at one of the **underlying** beliefs of systems planning: that of the planner as **superior**, scientific expert. From this, it was a short step to the notion that official participation in planning was itself a token action, designed to **manipulate** the public even further by offering them the shadow rather than the substance. In this view, what was needed was far more than mere **consultation** of the public; it was actual involvement of the citizenry in making plans for themselves. This was most appealing, but also evidently most difficult, in **deprived** urban areas where people were most **apathetic** and least well informed about the possibilities open to them.

The idea of community action in planning started in the United States, but spread rapidly to Britain in the **ferment** of ideas in the late 1960s, helped by the fact that at this time, there was a new concern with problems of social **deprivation** in the inner cities of both countries. From the start, it tended towards a radical **critique** of society and—especially in Britain—became heavily influenced by the intellectual currents towards Marxism at the time. This was perhaps predictable: community action depended on the idea that local people should be organized, and by definition this could not be done through the agency of **officialdom**; the people who set themselves up in this role were almost bound to believe in some radical mission to raise the people's consciousness. In the officially sponsored community development projects between 1972 and 1977, it rapidly led to **conflict** between the teams and the local councils, and so to the rapid **demise** of the experiment. But elsewhere, in a thousand different ways, it began to **generate** a great variety of semi-official and unofficial groups involved in various projects, with a wide variety of political views, from **liberal** left to Marxist left. Many of these came to play an important role in the inner cities when, after 1977, the government released funds for partnership and program authorities.

Marxism by then, however, was beginning to create a rather different **paradigm** of planning; it **overlapped** with community action in a number of places and in the behavior of a number of people, but for the most part it was rather distinct. It has come to be known as the political economy approach. Its essence is this: application of Marxist theory to the development of the modern capitalist economy reveals that very striking changes are taking place in the character of the economy of the advanced industrial countries of Europe and North America, and these are in turn having strong regional and urban impacts. In particular, **rationalization** of production is leading to major locational shifts of industry and to big reductions in workforces, which especially affect the older, bigger inner cities in the older industrial regions. The essence of this approach lies in analyzing the changes that are occurring, and the structural changes in ownership and control that **underlie** them. So far the major achievements of this school lie in analysis rather than either **prescription** or **proscription**—or, to put it another way, in urban studies rather than in urban planning. In so far as there have been policy recommendations, they tend to have been rather conventional ones in the form of an extension of the state sector, a growth of cooperative forms of production, and a control on the freedom of private industrial complexes to shut down plants. But underlying the whole analysis is a profound sense of the power of modern **multinational**, multi-plant corporations to affect the fortunes of cities and regions—a power that often seems far greater than the capacity of governments to influence their actions. In the event, this tradition of analysis proved **pervasive** during the 1980s, at a time when in both Britain and the United States right-wing governments were **retreating** from planning but encouraging development-led approaches to urban **regeneration**. And this led to a curious **divorce** between the theory and the practice in urban planning and development, which had never previously occurred.

The central problem with the neo-Marxist approach to planning of the 1970s and 1980s, **oddly**, seems similar to the problem of the systems planners whom the Marxists criticize. The burden of the Marxist critique is that the systems planners, claiming to be value free, never realized just how value dependent they were; they were mere technical planners who could discuss how to reach given ends, not the ends themselves. Only Marxists, whose training has allowed them to understand the laws of human social development, could pass through this **subtle veil**. But once they have achieved this, presumably they—like the systems planners before them—can **legitimately** claim to plan and to control. The problem, for them as for the systems planners, is why anyone should **heed** their claim to unique wisdom. The problem as to the **legitimacy** of planning remains; and, as a progressively larger section of the public becomes interested in the impacts of planning, it becomes more **acute**. Whatever the planners' ideology, it appears that people are no longer willing, as once apparently they were, to accept their claim to **omniscience** and **omnipotence**.

One answer is to help an increasingly well-informed and well-organized and active population to conduct better debate. That is why many of the most interesting developments in planning theory in the 1980s and 1990s have been about a kind of **transactive** planning. There were beginnings of this in the 1960s, in both the United States and Britain, but it has now become a much more **sophisticated** process, informed by a good deal of philosophical **underpinning** that owes much to the Marxist debates of the 1970s but goes beyond them, to try to strip away levels of false understanding and false representation, and that recognizes the essential complexity of many decisions and of the machinery necessary to resolve them.

Meanwhile, just because people are conscious that planning is a public good that can have both positive and negative impacts on them, **controversies** over planning proposals tend, if anything, to become more **vigorous** and even more **rancorous**. In this, it is not possible to argue simply that people are fighting the planners; often it appears—as over the line of a motorway or the location of a power station—that the people are fighting each other. Especially in periods of negative growth when there are all too few goods to go around, planning may become—in the words of the American economist Lester Thurow—a "zero–sum game": one in which if I win, you lose.

Whether zero–sum or not, few would doubt today that planning decisions critically affect what is known in the **jargon** as "real income". This, of course, is far more than money income: it includes such **intangible psychic** income as is provided by clean air, lack of noise, agreeable neighbors, freedom from crime, good education, a range of services **accessible** by efficient transport, and a **host** of other things. One important school of urban planners, therefore, regards questions of real income distribution as central to the planning process. Plan evaluation, in these planners' view, should be concerned less with **aggregate excess** of benefits over costs than with redistribution of real income so as to benefit the groups that now have the least. As I have already suggested, in this regard aggregate cost–benefit analysis is **inferior** to disaggregated analysis of the Planning Balance Sheet or Goals Achievement Matrix varieties. The latter approaches have the benefit that they specifically look at the distribution of costs and benefits, and the Planning Balance Sheet specifically looks as their **incidence** in different groups of the population. Additionally, they are capable of incorporating elements that cannot be accurately **rendered** in money terms, but that nevertheless form an important element in real income, such as gains and losses in environmental quality. This was one of the important new emphases of planning in the 1970s.

Parallel to it, and owing something to the political economy school and to the distributionist school, is an emphasis on generating economic growth. This, of course, reflected the concerns of the late 1970s and early 1980s. In **depressed** regions such as central Scotland, or Appalachia, growth was always a central concern; but by 1981 or 1982 it became all but **universal**. *Though there were strong **ideological** battles between right–wing and left–wing approaches on the question—the first favoring non–planning, land development–led approaches, the other stressing local authority–led schemes to **regenerate** traditional industrial enterprise—there was an **implicit** agreement on the **primacy** of*

revitalizing decayed *urban–industrial economies.* The **paradox** was that at the same time, the environmental concerns of the 1970s remained strong and that almost inevitably they **clashed** with the aim of economic regeneration. The clash was starkly highlighted in the United States, where environmental groups battled with government over issues like oil exploration off the California coast, or **strip mining** in western mountain states; but in Britain it is **illustrated** in subtler ways in the arguments about urban enterprise zones, or about mining rights in national parks.

The nature of the paradox is that this is a zero–sum society, but that to get out of that state, some groups would have to sacrifice something that they hold dear. Planning, in other words, is merely an **acute** instance of the central problem of society in the 1980s. And in the economic **revival** of the middle and late 1980s these **Nimby**–style issues became more and more prominent, and local pressure groups in the more favored areas sought to **erect** barriers to further growth in their areas.

The question, however, finally comes back to this: what, then, is the methodology of planning? How does it seek to resolve such a set of major problems? The answer should surely be: by some variant of the systems approach. It should not claim the instant ability to solve complex problems. It should not even necessarily claim unique expertise. It should certainly not claim to know what is good for people. Rather, it should be **exploratory** and instructive. It should aim to help communities think clearly and logically about resolving their problems, and in particular some of the more subtle underlying issues that concern such matters as **equity** or growth. It should try to examine alternative courses of action and trace through, as far as possible, the consequences of each of these for different groups of people in different places. It should not seek to avoid the difficult questions of who exercises political power on behalf of whom, and by what **legitimacy**. It should make recommendations, but it should not seek to **impose prescriptions**. It should claim modestly that planners may perhaps be more capable than the average person of conducting this kind of analysis, but not that they are uniquely expert. In other words, it should aim to provide a resource for democratic and informed decision–making. This is all planning can **legitimately** do, and all it can **pretend** to do. Properly understood, this is the real message of the systems revolution in planning and its aftermath.

Notes

[1] 本文选自：Hall, P. Urban and Regional Planning [M]. 4th edition. London and New York: Routledge, 2002. 彼得·霍尔.城市和区域规划（原著第四版）[M]. 邹德慈，李浩，陈熳莎，译.北京：中

国建筑工业出版社，2008。

[2] 主要相关文献：Castells, 1977; Chadwick, 1971; Faludi, 1973; Forester, 1993, 1999; Friedmann, 1988; Hall, 1988; Harvey, 1973; Healey, 1988, 1997; McLoughlin, 1969。

[3] 课文中下划线部分为有关城市规划理论的重要论述，需强化阅读，每篇课文习题部分均要求将下划线句子或段落译成中文。

[4] 课文中下划线斜体字部分为复杂长句，其解析详见下文“**Complex Sentences**”。

New Words

accessible adj. 易达到的；易受影响的〖前缀 3. ac–; 词根 30. cess〗

acute adj. 敏锐的，灵敏的；(病) 急性的〖词根 1. acu〗

aggregate adj. 聚集的，合计的；n. & v. 总数，合计〖前缀 3. ag–; 词根 86. greg〗

apathetic adj. 无感情的〖前缀 1. a–; 词根 153. path〗

clash v. & n. 冲突

conflict n. 冲突，战斗〖前缀 16. con–; 词根 73. flict〗

constrict v. 压缩，压紧；妨害，阻碍

consultation n. 商议；咨询

controversy n. 公开辩论，论战〖前缀 17. contro–; 词根 226. vers〗

critique v. & n. 批判，发表评论〖词根 48. crit〗

decay v. 腐烂，腐朽

decentralization n. 分散〖前缀 18. de–; 词根 32. centr〗

demise n. 死亡

depressed adj. 情绪低落的，沮丧的；萧条的〖前缀 18. de–; 词根 172. press〗

deprivation n. 剥夺，丧失

deprived adj. 缺乏食物的，缺乏足够教育的

disinterested adj. 公正的，客观的

divorce n. 离婚，分离

endorse v. 赞同，支持，认可

equally adv. 平等地，公正地〖词根 58. equ〗

equity n. 公平，公道〖词根 58. equ〗

erect v. 使直立，使竖起；建立，创立〖词根 177. rect〗

excess n. 过分，过度〖前缀 26. ex–; 词根 30. cess〗

exploratory adj. 探索的，考察的

ferment n. 酵素，发酵剂；骚动，动乱

generate v. 产生，引起〖词根 81. gene〗

heed n. & v. 注意，留心

homogeneous adj. 同类的，相似的〖前缀 30. homo–; 词根 81. gene〗

host n. 主人，东道主；寄主，宿主

ideological adj. 思想的，意识形态的

illustrate v. 说明，表明〖词根 117. lustr〗

implicit adj. 含蓄的〖前缀 33. im–; 词根 169. pli〗

impose v. 强加；征税〖前缀 33. im–; 词根 170. pos〗

incidence n. 发生，发生率；影响

incorporate v. 合并，并入〖前缀 34. in–; 词根 42. corpor〗

inferior adj. 低劣的，下级的，下等的

inflexible adj. 僵硬的，不可弯曲的〖前缀 34. in–; 词根 72. flex〗

intangible adj. 触不到的，难以理解的

jargon n. 行话，行业术语；黑话

legitimacy n. 合理，合法〖词根 105. leg〗

legitimately adv. 合法地，合理地〖词根 105. leg〗

liberal adj. 自由的，慷慨的〖词根 108. liber〗

manipulate v. 操作，处理〖词根 119. man〗

maximize v. 最大化，使增至最大限度

multinational adj. 多国的，跨国的〖前缀 43. multi–〗

Nimby, NIMBY n. 宁闭（指反对在本社区兴建有害设施的抗议者）(not in my back yard 的首字母缩合）

oddly adv. 奇怪地，古怪地

officialdom n. 官场，官僚作风

omnipotence n. 全能，万能，无限威力〖前缀 46. omni–; 词根 171. pot〗

omniscience n. 全知，全知者，上帝〖前缀 46. omni–; 词根 180. sci〗

overlap v. 重叠〖前缀 48. over–〗

paradigm n. 范例，样式

paradox n. 矛盾

participation n. 参与，参加

pervasive adj. 普遍的，渗透的，弥漫的

prescription n. 处方；训令，指示；法规

pressing adj. 紧迫的，紧急的〖词根 172. press〗

pretend v. 假装，装扮；自命，自称

primacy n. 首位，第一位〖词根 173. prim〗

proscription n. 禁止，剥夺权利

psychic adj. 精神的，灵魂的，超自然的

rancorous adj. 怨恨的，满怀恶意的

rationalization n. 合理化

react v. 反应，作出反应〖词根 2. act〗

reaction n. 反应〖词根 2. act〗

regenerate v. 使再生〖前缀 61. re–; 词根 81. gene〗

regeneration n. 再生，恢复〖前缀 61. re–; 词根 81. gene〗

render v. 使成为；递交；给予；表达

retreat v. 撤退，后退〖前缀 61. re–〗

revitalize v. 使新生，使复兴〖前缀 61. re–; 词根 230. vit〗

revival n. 复活，再生〖前缀 61. re–; 词根 231. viv〗

sophisticated adj. 复杂的，富有经验的

stagnant adj. 停滞的，不流动的

statutory adj. 法定的，法令的

strip mining 露天采矿

subsequently adv. 其后，随后，接着〖前缀 65. sub–; 词根 187. sequ〗

subtle adj. 微妙的，敏感的

superior adj. 较高的，较好的，较多的〖前缀 68. super–〗

tackle v. 着手处理，对付，解决

tenet n. 原则，信条，教义

transact v. 处理，办理；商议，谈判〖前缀 71. trans–; 词根 2. act〗

transactive adj. 商议的，谈判的〖前缀 71. trans–; 词根 2. act〗

underlie v. 位于……之下，构成……的基础〖前缀 75. under–〗

underlying adj. 潜在的，含蓄的；基础的，表面下的，下层的〖前缀 75. under–〗

underpinning n. 基础〖前缀 75. under–〗

universal adj. 普遍的，一般的，通用的〖前缀 76. uni–〗

veil n. 面纱；掩饰

vigorous adj. 有力的，精力充沛的

Complex Sentences

[1] In particular, **some planners began to question the basic tenets of the systems approach**: the notion **that** it was scientific, **in the sense that** the world could be completely understood and its future states predicted; the notion **that** planning could be value free, **in that** the planner could disinterestedly determine what was best for society; the notion **that** the planner was planning for

a society **that** was a homogeneous aggregate, in **which** the welfare of the entire people was to be maximized, **without** too much concern with distributional questions; and the notion **that** the task of planning was to come to terms with—**which**, in practice, mean adapting to—the facts of rapid growth and change.

黑体字部分为句子的主干(主语、谓语和宾语)。冒号后为宾语 the basic tenets 的四个同位语(四个 the notion),注意同位语结构复杂:第一同位语包括一个同位语从句(that 引导)和一个状语从句(in the sense that 引导);第二同位语包括一个同位语从句(that 引导)和一个状语从句(in that 引导);第三同位语包括一个同位语从句(that 引导),两个定语从句(that 和 which 引导)以及一个介词短语状语(without 引导);第四同位语包括一个同位语从句(that 引导),一个定语从句(which 引导)。

[2] Though there were strong ideological battles between right-wing and left-wing approaches on the question—the first favoring non-planning, land development-led approaches, the other stressing local authority-led schemes to regenerate traditional industrial enterprise—**there was an implicit agreement on the primacy of revitalizing decayed urban-industrial economies.**

黑体字部分为主句,though 引导的从句为让步状语从句,两个破折号之间部分为解释性插入语,对让步状语从句作进一步说明。

Exercises

[1] 将课文中下划线句子或段落译成中文。

[2] 读熟或背诵上述复杂长句。

[3] 解析下列单词[解析单词是指根据单词中包含的词根、前缀推导出单词的含义,例如,apathy(n. 漠然,冷淡),解析式为:a-(无)+ path(感情)+ y →无感情→冷漠,冷淡]。解析的单词选自上文"**New Words**"。

equity n. 公平,公道	inflexible adj. 僵硬的,不可弯曲的
erect v. 使直立,使竖起;建立,创立	liberal adj. 自由的,慷慨的
excess n. 过分,过度	omnipotence n. 全能,万能,无限威力
homogeneous adj. 同类的,相似的	omniscience n. 全知,全知者,上帝
illustrate v. 说明,表明	primacy n. 首位,第一位
implicit adj. 含蓄的	revival n. 复活,再生
impose v. 强加;征税	subsequently adv. 其后,随后,接着

Unit 14

Paradigm Shifts, Modernism and Postmodernism (1)

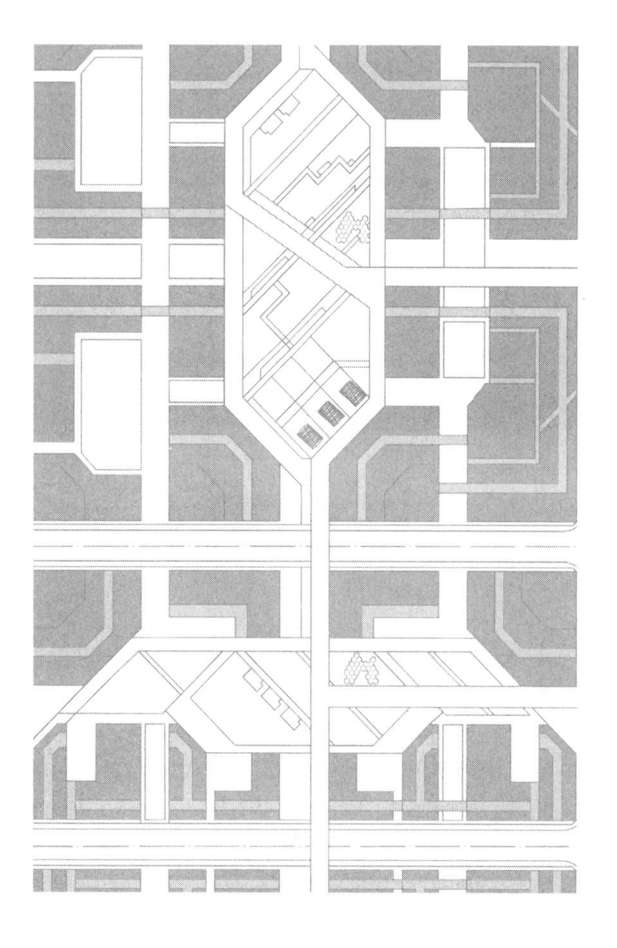

1 Changes in planning thought and paradigms

It will be clear from this book that ideas about town planning have changed over the fifty year period since the end of the Second World War. But what have been the most significant changes ? In this concluding chapter I offer a **retrospective** overview of the **evolution** of town planning thought since 1945, and in doing so, I shall try to describe the most significant shifts in planning thought over this period.

At various times since 1945 overviews of post-war planning theory have appeared. Galloway and Mahayni (1977) speak of "**paradigm** change" in post-war planning theory, and this term has come to be used widely by social theorists to describe major shifts in the history of ideas. It may therefore be useful to say something about the idea of "paradigms" to begin with, and how this **notion** might relate to the recent history of town planning theory.

As used to describe changes in thought, the idea of a "paradigm" **derives** from the work of Thomas Kuhn (1962), who employed the concept to describe major shifts in theoretical **perspective** in the history of science. According to Kuhn, if we look at the history of science, we find that advances in scientific thought have rarely occurred in a steady, evolutionary manner in response to the gradual **accumulation** of empirical evidence. Rather, the history of science is marked by long periods in which a given theoretical perspective—or "paradigm"—has prevailed and been accepted by members of a scientific community. During these relatively stable periods, most scientific research is **premised** upon the prevailing paradigm, and empirical observations are interpreted in terms of it. However, there is often some empirical evidence which does not "fit" **neatly** with the prevailing theoretical claims. Many scientists seem willing to "turn a blind eye" to this evidence on the assumption that one day someone will explain how it fits within the framework of the current paradigm. However, truly creative scientists are those who develop a new theoretical framework that succeeds in accounting for the hitherto puzzling evidence as well as the evidence previously explained by the "old" paradigm. When a new paradigm succeeds in **replacing** an old paradigm in this way, there is a revolution in scientific thought. For a whole way of perceiving and explaining some aspect of the world is **overthrown** and replaced by a new theoretical perspective. *As an example, the change from viewing the Earth as flat and at the centre of the Universe, to seeing it as round and as orbiting the Sun, was obviously a **profound** and revolutionary change in scientific thought–a paradigm shift in Kuhn's terms.* Another example given by Kuhn is the shift in the 20th century from the Newtonian to Einsteinian view of space and time (Kuhn, 1962). Once a new paradigm has become accepted, most scientific research comes to operate within this theoretical framework and, typically, another **quiescent** period of developing and refining this recently established theoretical framework **ensues**. So, according to Kuhn, the history of science continues.

It is clear from this account that, for Kuhn, paradigm changes are fundamental shifts in people's view of the world; that is why he describes paradigm shifts as revolutionary. And because they represent such fundamental changes in world view, such **paradigmatic** shifts typically occur **infrequently** in the history of science. Any given paradigm, once established, shapes the whole way a scientific community (and beyond that, the general public) views some aspect of the world, and tends to **endure** for centuries, not just decades.

We should therefore be cautious in applying the notion of paradigms, and paradigm shifts, to the changes in town planning thought which have occurred over the relatively short period of fifty years described in this book. We also need to note that Kuhn was describing changes in scientific thought; that is, major changes in the way people have described and explained some aspect of reality as a matter of fact. Town planning is not, in the strict sense, a science (not even—as some still persist in saying—a social science). Rather, it is a form of social action, directed at shaping the physical environment, and driven in this by certain moral, political, and aesthetic values. In other words, town planning is an "ethical" (and hence political) practice, though of course, in seeking to realize certain valued ends, town planning should draw on relevant scientific understanding.

Nevertheless, providing we are **alert** to the dangers of overusing Kuhn's concept of paradigm shifts there is nothing to stop us using it in a looser or more generous way to describe significant changes in town planning thought. Moreover, the notion of paradigm shifts may also be applied to fundamental shifts in values or ethical thinking. With these **precautions** in mind, I shall explore the **appropriateness** of applying the Kuhnian **terminology** of paradigm shifts to describe the main changes in town planning theory since 1945.

This overview of post-war planning theory is organized as follows. In the next section I shall offer a view of the two most significant changes in town planning thought since 1945. The first of these occurred in the 1960s, with the shift from the urban design tradition to the systems and rational process views. The second change occurred during the 1970s and 1980s, and represented a shift in view of the planner's role. *In particular, there was a shift from a view of the planner as a technical expert to the view of the planner as a kind of "**facilitator**", drawing in other people's views and skills to the business of making planning judgements.*

I then evaluate post-war planning thought from a more general perspective, situating changes in planning thought within the larger context of general changes in **contemporary** thought and culture. In particular, I describe the thesis put forward by some writers that from the late 1960s to the 1980s, a significant change in the history of western thought occurred from what has been called "modernism" to "postmodernism". Arguably, this more general, cultural change in thought and values has also had a significant impact on town planning thought.

I conclude the chapter with some final **reflections** on town planning as a discipline, and the kind of theory that is relevant to this discipline.

2 Two paradigm shifts in planning thought since 1945?

2.1 From town planning as design to science

Chapter 1 described how, for almost 20 years following the Second World War, British town planning theory and practice was dominated by a concept which saw town planning essentially as an exercise in physical design. Its long historical **lineage** is shown by the fact that, for as far back as we can see, what came to be seen as town planning was assumed to be most appropriately carried out by architects. Indeed, such was the **intimate** connection between architecture and town planning that the two were not distinguished throughout most of human history. Thus what we call town planning was seen as architecture; its only distinctiveness being that it was architecture on the larger scale of a whole town, or at least part of a town, as distinct from an individual building.

The concept of town planning as "architecture **writ large**" persisted to the 1960s, as is shown by the fact that most planners in the post-war years were architects by training, or "architect–planners". Like architecture, town planning was viewed as an "art", albeit (again like architecture) an "applied" or "practical" art in which utilitarian or "functional" requirements had to be **accommodated**. *Hence the systems and rational process views of planning which burst on to the scene in the 1960s represented a **rupture** with tradition—a change in planning thought which can be seen as a paradigm shift in the most fundamental, Kuhnian sense.*

In Chapter 4 I stressed that the systems and rational process views of planning were conceptually distinct and thus really two theories of planning. Thus the systems view was based on a view about the object that town planning deals with (the town, or the environment in general, was viewed as a "system"), whereas the rational process view concerned the process of planning itself. But both views, taken together, represented a **departure** from the prevailing design–based view of town planning. This shift in planning thought can be summarized under four points.

First, an essentially physical or **morphological** view of towns was replaced with a view of towns as systems of inter–related activities in a constant state of **flux**. Secondly, whereas town planners had viewed and judged towns **predominantly** in physical and aesthetic terms, they now examined them in terms of social life and economic activities; in Harvey's (1973) terms, a sociological conception of space replaced a geographical or morphological conception of space. Thirdly, because the town was now seen as a "live" functioning thing, this **implied** a "process" rather than an "end–state" or "blueprint" approach. Fourthly, all these changes implied in turn a change in the kinds of skills appropriate to town planning. If town planners were trying to control and plan complex, **dynamic** systems, what seemed to be required were **rigorous** "scientific" methods of analysis. Overall, the shift in planning thought suggested that town planning was a science, not an art. For the analysis of environmental systems involved systematic empirical investigation, while the concept of planning as a process of rational decision–making seemed to make planning a "scientific" exercise because "rationality" was **equated** with science. This shift was so significant that it was **profoundly unsettling** to many planners and planning students **reared** in the design tradition of town planning. *Suddenly, town planners who had seen themselves as "artistic" urban designers were being told by a new generation of planning theorists that their former conception of town planning was **inappropriate** and that they should see themselves as "scientific" systems analysts.*

However it is important to note that the design–based tradition of town planning was not completely **superseded** by the theoretical changes of the 1960s. Although questions of design and aesthetics were **marginalized** in planning theory for over 20 years, in practice the development control sections of local planning authorities continued to assess development proposals in terms of their design and aesthetic impact. Moreover, the well-known Essex Design Guide (Essex County Council Planning Department, 1973), was the first sign that many local authorities were seeking to place their practice of design control on a clearer theoretical footing by **articulating** "principles" of good design. This was significant. For it drew attention to the fact that, at the level of "local" planning, the physical form and aesthetic appearance of new development remained a necessary and significant consideration for town planning. It was therefore only at the broader, more strategic level of planning that the design-based view of planning was **supplanted** by the changing conception of planning **ushered in** by the systems and rational process views of planning.

The revolution in town planning thought of the 1960s did not involve the complete replacement of one view of town planning by another. The real revolution was in making a distinction between two levels of town planning, one strategic and longer term, and the other "local" and more immediate. The altered concept of town planning brought about by systems and rational process thinking was most appropriate at the strategic scale, although there were also lessons for "local" small area planning in systems and rational process thinking (e.g. in giving greater consideration to social and economic factors, in approaching local planning as a rational process, etc.) . In **retrospect**, then, the revolution in town planning thought in the 1960s was not a whole-scale revolution which completely **ousted** the **incumbent** design-based view of town planning. Seen like this the **advent** of the systems and rational process views of planning did not represent a paradigm shift of the most fundamental kind described by Kuhn.

2.2 From the town planner as technical expert to "communicator"

Although the shift from seeing town planning as an exercise in physical planning and design to seeing it as a rational process of decision-making directed at the analysis and control of urban systems was undoubtedly a radical shift in thinking, there was one thing that the design-based view, and the systems and rational process views, had in common. It was that the town planner was someone who possessed some specialist skill—some expertise—which the average person in the street did not possess. It was this which qualified the planner, **literally**, to plan. And it was this, too, which made town planning a distinct "profession" (the possession of some specialist knowledge or skill being the **hallmark** of any claim to professional status).

With such changes to planning thought there were **inevitably** changes in views about the specialist skills a planner required. The traditional design-based view of town planning demanded skills of aesthetic appreciation and urban design but the systems and rational process views demanded those of scientific and logical analysis. However, in the 1960s it was also acknowledged that town planning judgements were at root judgements of value (as distinct from purely technical judgements) about the kinds of environments it is desireable to create or **conserve**. The question therefore arose as to whether town planners had any greater "specialist" ability to make these judgements than the ordinary person in the street. The experience of much 1960s planning—such as comprehensive housing redevelopment or urban road planning—seemed to indicate not. The view that town planning was a **value-laden**, political process therefore raised not so much the question of what the town planner's area of specialist expertise should be but, more fundamentally, the question of whether there was any such expertise at all.

From this questioning developed a **curious bifurcation** in planning theory which has persisted to this day. On the one hand, some planning theorists have continued to believe that the practice of town planning requires some specialist substantive knowledge or skills—be it about urban design, urban regeneration, sustainable development or whatever. On the other has developed a tradition of planning thought which openly acknowledges that town planning judgements are value-laden and political. One conclusion would be to reject entirely the idea that town planning involves some special expertise at all, and some "radical" planning theorists have flirted with this view (e.g. Goodman, 1972; Evans, 1995). However, most planning theorists who have openly acknowledged the value-laden and political nature of planning have developed an **alternative** line of thought which rejects the idea that the town planner is someone who is specially qualified to make better decisions. What is "better" is a matter of value, and planners have no **superior** expertise in making value-judgements about environmental options. However, the view is still taken that the town planner possesses some specialist skill, namely, skill in managing the process of arriving at planning decisions. A tradition of planning theory has emerged, therefore, which views the town planner's role as one of identifying and **mediating** between different interest groups involved in land development. The town planner is seen as someone who acts as a kind of cypher for other people's assessments of planning issues, rather than someone who is better qualified to assess these issues him or herself. The town planner is viewed as not so much a technical expert (i.e. as someone who possesses some superior skill to plan towns), but more as a "**facilitator**" of other people's views about how a town, or part of a town, should be planned.

An early version of this theory was Paul Davidoff's "**advocacy**" view of planning (see Chapter 5). *The most recent version is the communicative planning theory described in*

*Chapter 7, under which the skills of **interpersonal** communication and negotiation are seen to be central to a **non-coercive**, "facilitator" model of town planning.* It has even been suggested that the kinds of interpersonal skills needed by the communicative town planner are those of the listener and the **counsellor**:

> Meaningful dialogue—learning the language of the **client**—is at the heart of effective counselling. To counsel is not to give advice or push the client down a particular path, but to let the client see himself or herself fully and through this discovery achieve personal growth. As local government offices look for ways of including citizens in decision-making, they must adopt many counselling skills—active listening, non-judgemental acceptance, and the ability to **empathize**. How can people play a part in the decision-making process unless we 'enable' them to do so? (Healey and Gilroy, 1990).

This is a far cry from the view that the specialist skill of the town planner **resides** in being either an urban designer or a systems analyst.

However, once again a word of caution is in order before it is too readily assumed that the difference between the two theories represents something like a paradigm shift, for it is possible to imagine some kind of **merger** between the two views. Thus a view of the town planner as primarily a communicator and negotiator can **accommodate** the planner having specialist knowledge which, for example, would enable him or her to point out the likely consequences of development proposals on the form and functioning of a town. Such a model of the town planner would be **akin** to that of, say, the civil servant who is an expert in economic matters, and who **imparts** his specialist economic understanding to those he advises who make decisions about economic policy. To be effective as an adviser, such a town planner would have to be skilled in communicating and negotiating with others, but he would also have to possess some specialist knowledge to bring to the communicating table to assist others in arriving at planning decisions.

Notes

[1] 本文选自：Taylor, N. *Urban Planning Theory since 1945* [M]. Thousand Oaks: Sage. 尼格尔・泰勒，1998. 1945 年后西方城市规划理论的流变 [M]. 李白玉，陈贞，译 . 北京：中国建筑工业出版社，2006。

[2] 本文主要相关文献：Alexander, 1965; Davidoff, 1965; Faludi, 1973b; Galloway, et al, 1977; Giddens,

1990; Harvey, 1973; Innes, 1995; Jacobs, 1961; Jencks, 1991; Kuhn, 1962; Sager, 1994; Venturi, 1966。

[3] 课文中下划线部分为有关城市规划理论的重要论述，需强化阅读，每篇课文习题部分均要求将下划线句子或段落译成中文。

[4] 课文中下划线斜体字部分为复杂长句，其解析详见下文“**Complex Sentences**”。

New Words

accommodate v. 容纳，向……提供；使适应，使和谐一致

accumulate n. 堆积，积累〖前缀 3. ac–; 词根 51. cumul〗

advent n. 出现，到来

advocacy n. 拥护，支持〖前缀 3. ad–; 词根 232. voc, vok〗

akin adj. 同族的，相似的

alert adj. 警觉的，警惕的

alternative adj. 择一的，供选择的；n. 选择，替换物〖词根 7. alter〗

appropriate adj. 适当的，恰当的

articulate adj. 发音清晰的，善于表达的；v. 清晰地发音，明确表达

bifurcation n. 分歧〖前缀 12. bi–〗

client n. 委托人，当事人，顾客

conserve v. 保护，保存，保藏〖前缀 16. con–; 词根 188. serv〗

contemporary adj. 当代的，同时代的〖前缀 16. con–; 词根 207. tempor〗

counselor n. 顾问，咨询员

curiously adv. 好奇地；奇怪地

depart v. 离开，起程，出发

derive v. 得到，导出；源于，来自

dynamic n. 动态，动力；adj. 动态的，动力的

empathize v. 表同情，有同感；使神入，使感情移入〖前缀 23. em–; 词根 153. path〗

endure v. 持续，持久

ensue v. 接踵发生，继而产生

equate v. 相等，等同〖词根 58. equ〗

evolution n. 发展，进化〖词根 234. volu〗

facilitator n. 促进者，帮助者

flux n. 流量，流出〖词根 75. flu〗

hallmark n. 特点，标志

impart v. 传授，告知

imply v. 暗示〖前缀 33. im–; 词根 169. pli〗

inappropriate adj. 不恰当的，不适宜的〖前缀 34. in–〗

incumbent adj. 在职的；n. 在职者〖前缀 34. in–; 词根 50. cumb〗

inevitably adv. 不可避免地，必然地〖前缀 34. in–〗

infrequently adv. 稀少地，珍贵地〖34. in–〗

interpersonal adj. 人与人之间的，人际的〖前缀 35. inter–〗

intimate adj. 亲密的，亲近的

lineage n. 血统，世系

literally adv. 逐字地，照字面地；确实地，真正地〖词根 110. liter〗

marginalize v. 使处于边缘，忽视

mediate v. 调解，调停，斡旋〖词根 122. medi〗

merge v. 混合，融入

morphological adj. 形态的，形态学的〖词根 131. morph〗

neatly adv. 整洁地，灵巧地，恰好地

non–coercive adj. 非强制的，非强迫的〖前缀 45. non–〗

notion n. 概念，观念〖词根 139. not〗

oust v. 驱逐，罢黜；取代

overthrow v. 打倒，推翻〖前缀 48. over–〗

paradigm n. 范例，样式

paradigmatic adj. 范例的

perspective n.（判断事物的）角度，方法；透视法〖前缀 53. per–; 词根 198. spec〗

precaution n. 预防，防备〖前缀 57. pre–〗

predominantly adv. 显著地，占优势地

premise n. 前提；v. 以……为前提

profound adj. 深厚的，深远的

quiescent adj. 静态的，不活动的〖词根 176. qui〗

rear v. 养育，抚养，饲养

reflection n. 反映；思考

replace v. 把……放回原位；替换，代替〖前缀 61. re–〗

reside in（性质等）存在

retrospect n. & v. 回顾，回想〖前缀 62. retro–; 词根 198. spec〗

retrospective adj. 回顾的，怀旧的〖前缀 62. retro–; 词根 198. spec〗

rigorous adj. 严厉的，严格的，严酷的，严密的

rupture n. 断裂，破裂〖词根 178. rupt〗

superior adj. 较高的，较好的，较多的，上等的〖前缀 68. super-〗

supersede v. 取代，接替〖前缀 68. super-; 词根 183. sed〗

supplant v. 取代，代替〖前缀 67. sup-〗

terminology n. 专门名词，术语

unsettling adj. 使人不安的〖前缀 74. un-〗

usher in v. 预报……的到来；宣告，展示

value-laden adj. 承载价值的 (laden adj. 满载的，负载的)

writ large adj. 显然扩大或夸大了的，显而易见的

Complex Sentences

[1] As an example, **the change** from viewing the Earth as flat and at the centre of the Universe, to seeing it as round and as orbiting the Sun, **was obviously a profound and revolutionary change in scientific thought**—a paradigm shift in Kuhn's terms.

黑体字部分为句子的主干（主语、谓语和表语），from...to... 结构介词短语作定语，修饰其前的 the change, 破折号后面内容为解释性插入语。

[2] In particular, **there was a shift** from a view of the planner as a technical expert to the view of the planner as a kind of "facilitator", drawing in other people's views and skills to the business of making planning judgements.

黑体字部分为句子的主干，from...to... 结构介词短语作定语，修饰其前的 a shift, drawing 引导的现在分词短语作定语，修饰前面的 facilitator。

[3] Hence **the systems and rational process views of planning** which burst on to the scene in the 1960s **represented a rupture with tradition**—*a change in planning thought which can be seen as a paradigm shift in the most fundamental, Kuhnian sense.*

黑体字部分为句子的主干（主语、谓语和宾语），第一个 which 引导的从句为定语从句，修饰其前的 the systems and rational process views of planning，第二个 which 引导的从句亦为定语从句，修饰其前的 a change in planning thought, 破折号后面内容为解释性插入语。

[4] Suddenly, **town planners** who had seen themselves as "artistic" urban designers **were being told by a new generation of planning theorists** that their former conception of town planning was inappropriate and that they should see themselves as "scientific" systems analysts.

黑体字部分为句子的主干（主语、谓语和宾语，被动语态），who 引导的从句为定语从句，修饰前面的 town planners，两个 that 引导的从句为宾语从句。

[5] *Although **the shift** from seeing town planning as an exercise in physical planning and design to seeing it as a rational process of decision-making directed at the analysis and control of urban systems **was undoubtedly a radical shift in thinking**,* **there was one thing** that the design-based

view, and the systems and rational process views, had in common.

正体黑体字部分为句子的主干，其后 that 引导的从句为定语从句，修饰其前的 one thing。斜体字部分为 although 引导的让步状语从句，斜体黑体字部分为让步状语从句的主干，from...to... 结构介词短语作定语，修饰其前的 the shift, directed 引导的过去分词短语作定语，修饰前面的 a rational process of decision–making。

[6] **The most recent version is the communicative planning theory** described in Chapter 7, under which the skills of interpersonal communication and negotiation are seen to be central to a non–coercive, "facilitator" model of town planning.

黑体字部分为句子的主干，described 引导的过去分词短语作定语，which 引导的从句为定语从句，两者均修饰 the communicative planning theory。

Exercises

[1] 将课文中下划线句子或段落译成中文。

[2] 读熟或背诵上述复杂长句。

[3] 解析下列单词 [解析单词是指根据单词中包含的词根、前缀推导出单词的含义，例如，apathy（n. 漠然，冷淡），解析式为：a-（无）+ path（感情）+ y →无感情→冷漠，冷淡]。解析的单词选自上文"**New Words**"。

bifurcation n. 分歧	perspective n. 角度，方法；透视法
empathize v. 有同感，使感情移入	quiescent adj. 静态的，不活动的
equate v. 相等，等同	retrospective adj. 回顾的，怀旧的
imply v. 暗示	rupture n. 断裂，破裂
incumbent adj. 在职的；n. 在职者	superior adj. 较高的，较好的，上等的
mediate v. 调解，调停，斡旋	supersede v. 取代，接替
morphological adj. 形态的，形态学的	supplant v. 取代，代替

Unit 15

Paradigm Shifts, Modernism and Postmodernism (2)

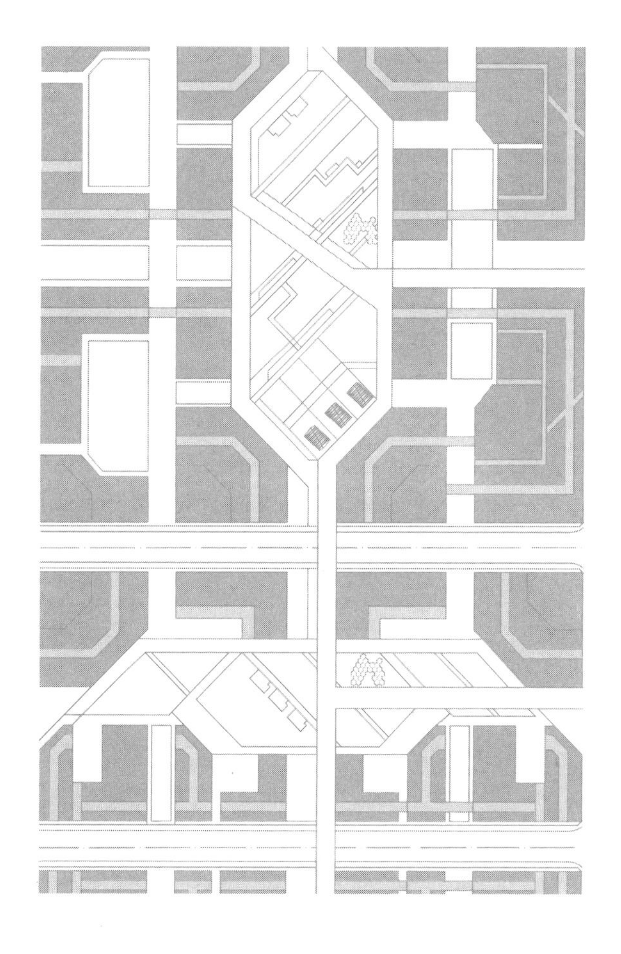

3 Modernism and postmodernism

According to some contemporary thinkers, over the last twenty years or so there has been a sea change in western thought and culture from "modernism" to "postmodernism"–a change so significant as to represent something ***akin*** *to a Kuhnian paradigm shift.* If there has been such a fundamental shift then town planning thought could hardly have escaped its impact. The **alleged** shift from modernism to postmodernism is particularly relevant to town planning as town planning and architecture have been one of the main "sites" where the shift from modernism to postmodernism has most clearly occurred. *Thus according to one of the leading writers about postmodernism, Charles Jencks (1991), the modern era ended when, on 15 July 1972, the Pruitt–Igoe housing estate in St Louis (USA)—which had earlier won an award as an exemplar of modern architecture and town planning—was* ***dynamited*** *and destroyed by the local city authority.*

Historical change is, of course, rarely as dramatic as this. The high–rise housing estates built all over Europe and North America have long been fiercely criticized as **soulless**, **inhumane** environments. The "functional" style of modern architecture was criticized even in its **infancy** for its **anonymity** and lack of visual interest (e.g. the 1920s debate between the Dutch de Stijl architects, who favored the **austere geometrical** forms which were to become the norm of modern architecture, and the **adherents** to the Amsterdam School of architecture, who favored an architecture of expressive forms and decorative brickwork). However, from the late 1960s onwards the rejection of modernism became stronger and more widespread and developed into the movement now called postmodernism.

3.1 Postmodernism as a style

At one level, postmodernism represented a **reaction** against the styles of art and design which had been **promoted** by the modern movement. Postmodernists rejected the **pared-down** simplicity of modern "functional" architecture, and so sought to "bring back style" to enrich the aesthetic content of contemporary buildings. Thus Robert Venturi (1966), in what is arguably the first text of postmodern architecture, famously **counterposed** his preference for a stylistically more complex architecture over plain "functional" modernism:

> I like complexity and **contradiction** in architecture...Architects can no longer afford to be **intimidated** by the **puritanically** moral language of **orthodox** Modern architecture. I like elements which are **hybrid** rather than "pure", compromising rather than "clean", **distorted** rather than "straightforward", **ambiguous** rather than "**articulated**", ...inconsistent and **equivocal** rather than direct and clear. I am for **messy vitality** over obvious unity...I am for richness of meaning rather than clarity of meaning.

A similar preference for complexity and "richness" over simplicity and order was voiced by Jane Jacobs for town planning in *The Death and Life of Great American Cities (1961)*. Jacobs **berated** modern city planning for its simplemindedness, as represented, for example, in the single-use zoning of urban areas, or in its **uncompromising** approach to "comprehensive" redevelopment which showed little understanding of the delicate social and economic **fabric** of so-called **slum** areas. Jacobs suggested that successful city areas were those with mixed uses, and that currently run-down "slum" areas could, if left alone by planners, naturally "un slum" themselves.

According to some accounts, however, the shift from modernism to postmodernism goes deeper than aesthetic preference. **Underpinning** the "modern" movement generally was a more fundamental intellectual **orientation** involving a reliance upon reason and science—an **optimistic** belief that, through rational analysis and greater scientific understanding, humans could create a better world for themselves. The technological urban **Utopias** produced by modern architects in the early years of the twentieth century were expressions of this modernist faith in science and technology. Such was the self-confident optimism of architects like Le Corbusier and the Italian Futurists that they **advocated** sweeping away the traditional city to make way for their modern urban Utopias. Sixty or so years later, following the horrors wrought by modern technology, including the horrors of the Corbusier experiment of the comprehensively planned high-rise city, this confidence in modern science and technology has been seriously **dente**d. Some postmodernists have offered a critique of the modernist reliance on science and even reason itself (e.g. Rorty, 1989; see also Truett Anderson, 1995). From this viewpoint, the **dynamiting** of the Pruitt-Igoe housing estate was not just an act to get rid of one housing form but it was also a symbol of a more fundamental **collapse** of the modernist faith in human rationality and scientific technology as the keys to a brighter human future.

3.2 Postmodernism as a challenge to science and reason

The modernist faith in reason and science had its roots in the European Enlightenment of the eighteenth century. What Habermas (1981) has called the "project" of modernity was really a technological development of the Enlightenment, and the idea that cities could be made better by rational thought and action—by "planning"—was thus part of this project. The postmodern critique therefore brings into question the Enlightenment assumption that the **deficiencies** of cities and our environment generally can be made better by planned action. Hence the real radicalness of Jacobs's critique was not just her preference for mixed uses over single use zoning, etc., but her **implicit** questioning of whether cities could be made better places by rational planning at all. This proposition is certainly a radical one, and again might suggest a paradigm shift in the most fundamental sense.

Some of the most radical versions of postmodernism therefore bring into question the **efficacy** of reason itself. Such a position is stated, for example, by Michael Dear (1995, cited in Healey et al., 1995):

> Postmodernism's principal target has been the rationality of the modern movement, especially its foundational character, its search for universal truth...The postmodern position is that all meta-narratives are suspect; that the authority claimed by any single explanation is **ill-founded**, and hence should be resisted. In essence, postmodernists assert that the relative merit of one meta-narrative over another is ultimately **undecidable**; and by **extension**, that any such attempts to **forge** intellectual **consensus** should be resisted.

Taken at face value this statement implies a rejection of rational **discourse** altogether. For if postmodernists believe, as Dear here suggests, that there are **literally** no criteria against which we can judge the relative merits of different theoretical positions, then it follows that there can be no reasoned debate about different theories at all. However, apart from the fact that such a position is intellectually hopeless (in the literal sense that there would be no point in hoping for greater enlightenment through rational discourse with others), it is also **self-defeating**. For if there are no rational criteria against which to judge the truth of a proposition, then there are no reasonable grounds for accepting as true the postmodern proposition that the "relative merit of one meta-narrative over another is ultimately undecidable". As Anthony Giddens (1990) has written: "Were anyone to hold such a view (and indeed if it is not **inchoate** in the first place), they could scarcely write a book about it. The only possibility would be to **repudiate** intellectual activity altogether."

In fact, many writers who advance an extreme "postmodernist" position of **epistemological** relativism are themselves **inconsistent** in **adhering** to it. Thus within a page of the position quoted above, we find Michael Dear writing about what the city of Los Angeles is "really like" on the assumption, presumably, that there are truths about the place which can be discovered. The idea, then, that there are no standards of rationality which we should **aspire** towards in engaging in theoretical work should be dismissed, as Giddens again puts it, "as unworthy of serious intellectual consideration".

The same applies to those who criticise the ideal of rationality in relation to town planning. By all means, we may reject as undesirable and unrealistic the **pursuit** of comprehensiveness in town planning, whether in relation to the actual practice of replanning cities comprehensively or in relation to the process of making planning decisions. As argued earlier in this book (Chapter 4), the pursuit of rationality is distinct from the question of whether we are required to be comprehensive. But the proposition that rationality does not matter in planning theory, or that a rational approach to the process of planning is no better than any other approach, is open to the same criticisms as the postmodernist rejection of rationality made above.

If these arguments are **persuasive**, then the idea that postmodernism represents a **paradigmatic** break with Enlightenment reason turns out to be empty. The employment of reason in planning, aided by the best possible scientific understanding of the world we are seeking to plan, remains as relevant and important now as it has ever been. As for the critique of "planning", it is important to recognize that what postmodernists have criticized is modernist planning (i.e. "clean-sweep" comprehensive planning), rather than necessarily planning as such. After all, there can be different styles of town planning, some of which are **compatible** with the ideals **articulated** by Jacobs. The idea that some cities have grown "naturally", as Alexander (1965) puts it, is **misleading**. For most human action is planned to some degree. We might talk of some cities having grown in more **piecemeal**, **incremental** and "**organic**" ways, but then it is possible to **envisage** styles of town planning which are precisely like this.

3.3 Postmodernism as an alternative normative view of the quality of life

The conclusion reached in the previous section does not necessarily **imply** a rejection of postmodernism entirely, for postmodernism also represents a shift in thinking about style and aesthetics, and it is here that the real significance of postmodernism lies. Postmodernism, however, does not just question certain styles but **posits** some alternative values of a more general kind—an alternative view, in fact, of the quality of life. These alternative values bring into question many of the values and **normative** principles which have **informed** town planning in the modern age, and it is from this point of view that postmodernism presents a case for the serious reconsideration of the purposes of town planning.

What, then, are these alternative postmodern values? Postmodernists argue that the world and our experience of it is far more complex and subtle than has typically been realized. Thus in relation to cities, postmodernists claim that people's experience of places, and from this the qualities of places, are much more **diverse** and "open" than was **implicit** in many modern schemes, and especially the **bombastic** simplicities of modern architectural visions of the ideal city. Instead of the modernist emphasis on simplicity, order, uniformity and tidiness, postmodernists typically **celebrate** complexity, **diversity**, difference, and **pluralism** (cf. Marion Young, 1990). Hence there can be no one type of environment which is ideal for everyone, no singular conception of environmental quality. Thus some may hold as an ideal Howard's **genteel** garden cities, but others will prefer the **buzz** and excitement of big city life–the "teeming metropolis" as Elizabeth Wilson calls it (1991).

Central to postmodernist values is a celebration of big city life because of its diversity and pluralism, and for the freedom of choice that this diversity promises. These values connect postmodernism with **liberalism**, for **liberals** also celebrate the **plural** society in which individuals have the opportunity to determine and "realize" themselves through the exercise of free choice. *It is these values which allow us to see Venturi (in relation to architecture) and Jacobs (in relation to town planning) as early postmodernists, for they argued for complexity and diversity as opposed to the* ***sterile*** *simplifications of modernism.*

All this is very general and, indeed, one of the criticisms which can be leveled at the normative position of postmodernism is that it is so general as to be **elusive**. The celebration of diversity, for example, can be taken to an extreme where anything that is different may be accepted or permitted; in other words, a position of moral and political **relativism** corresponding to, and equally as **untenable** as, the **epistemological** relativism of some postmodernists discussed earlier. Thus in town planning, might there not be some values and ideals which town planning should **aspire** to, wherever it is practised? For example, shouldn't town planning, everywhere, do what it can to help bring about economically and environmentally sustainable development, development which is not socially divisive, and development which is experienced as an aesthetic delight? (In other words, shouldn't town planning be broadly guided by the ideals which **underpin** the five areas of "problem-based" research and theory described in Chapter 8?) If so, then whatever postmodernists say about the virtues of diversity and pluralism, and however important this is as a lesson for town planning, there may still be some **overarching** "universal" ideals which town planning should aspire to.

4 Conclusion: town planning as a discipline and town planning theory

In conclusion, we come back to town planning theory. At various times since 1945 town planning theorists have written about planning theory as if there could be, or should be, only one kind of "planning theory". Faludi's (1973) early view that it was only theory about the planning process (procedural planning theory) which **warranted** the title "planning theory" was an example of this. We also see this intellectual **imperialism** in the later rejection of procedural planning theory, and the alternative suggestion that town planning theory should necessarily be empirically based theory about the role of planning within its political economic context (e.g. Scott and Roweis, 1977).

The truth is that there are different types or kinds of theories, answering different kinds of questions, and not only one type of theory is relevant to town planning. Scientific theories which seek to improve our understanding of the world that town planning is dealing with, including the effects of different kinds of planned actions on the world, are fundamental to sound town planning. But town planning exists to improve the world, not just to understand it. Therefore philosophical reflection on the purposes of planning, such as that which postmodernism has prompted, is also central to planning theory. In other words, normative theory—including moral and political philosophy—is also a proper part of town planning theory.

Any account of what theory is relevant to town planning **presupposes** some conception of what sort of an activity town planning is, and much more could be said about this than space here allows. However, two broad observations about the nature of town planning further **illuminate** the kind of theory most relevant to its practice.

First, town planning is a form of social action, or a social practice. It is about **intervening** in the world to protect or change it in some way—to make it other than it would otherwise be without planning. Because it is a practice, it requires, more than anything, sound judgement—judgement about what best to do. In this respect, theory about practical reasoning and judgement is absolutely central to town planning, as John Forester has insisted in developing his version of communicative planning theory. Seen thus, town planning is neither an art nor a science in the strict sense of either of these terms, though, of course, sound judgement in town planning draws on both aesthetic and scientific understanding.

Because town planning is a practical discipline, some town planners have been **skeptical** about the value of town planning theory, and in Britain this has generated a most unfortunate "anti-intellectualism" in the planning profession, as Reade (1987) has emphasized. Certainly, if a theory has no bearing on the practice of town planning, then there is no need for practising planners to consider it. But that is not an argument against theory, only an argument against **irrelevant** theory. And so, even if the rejection of certain kinds of theorising is justified, this does not justify the rejection of theory entirely. On the **contrary**, precisely because town planning is a practical discipline which directly affects the environment in which people live, it is all the more essential to develop good practical theory to **inform** it. And, as we saw from our analysis of early post-war planning theory, what some people think of as practical "common sense"' is often not good practical theory at all.

The second main point to note is that town planning involves making judgements about what best to do—that is, about how best to plan the environments we inhabit. Throughout this book attention has been drawn to the fact that town planning is fundamentally about making value-judgements about the kinds of environments we want to protect or create. It follows that, at the heart of town planning theory, there should be **rigorous** analysis of, and theories about, environmental quality: what constitutes it; what different views (if any) different groups take of it; what different sorts of qualities (e.g. economic, social, aesthetic, ecological) make up quality environments and what possible **tensions** can arise between these different components of environmental quality; and how good-quality environments have been created in the past, and how they are most likely to be created in the conditions in which we now find ourselves.

One final thought. Of the two areas of practical and normative theoretical inquiry described above, we have, perhaps, got further with the former than with the latter. That is, since 1945, we have learnt more about how best to go about the practical process of town planning than we have learnt about the kinds of environmental qualities town planning should be aiming at. For one thing which **emerges** from this account is that considerably more theoretical attention has been devoted to refining our conception of what kind of an activity town planning is, and from this how best it should be approached, than has been devoted to analyzing the **constituents** of high-quality environments and how they might be realized. *If this is true, then it suggests that one of the most important tasks facing town planning theorists now is the development of better theory about the environmental qualities which town planning practice should help bring into being.*

Notes

[1] 本文选自：Taylor，N. *Urban Planning Theory since 1945*[M]. Thousand Oaks：Sage. 尼格尔・泰勒，1998. 1945 年后西方城市规划理论的流变 [M]. 李白玉，陈贞，译 . 北京：中国建筑工业出版社，2006。

[2] 本文主要相关文献：Alexander, 1965; Davidoff, 1965; Faludi, 1973b; Galloway, et al, 1977; Giddens, 1990; Harvey, 1973; Innes, 1995; Jacobs, 1961; Jencks, 1991; Kuhn, 1962; Sager, 1994; Venturi, 1966。

[3] 课文中下划线部分为有关城市规划理论的重要论述，需强化阅读，每篇课文习题部分均要求将下划线句子或段落译成中文。

[4] 课文中下划线斜体字部分为复杂长句，其解析详见下文“**Complex Sentences**”。

New Words

adhere v. 黏附；坚持〖前缀 3. ad–; 词根 90. here〗
adherent n. 支持者，拥护者〖前缀 3. ad–; 词根 90. here〗
advocate v. 拥护，支持，鼓吹；n. 支持者，拥护者〖前缀 3. ad–; 词根 232. voc〗
akin adj. 同族的，相似的
allege v. 陈述，宣称
ambiguous adj. 含糊的，模棱两可的〖前缀 5. ambi–〗
anonymity n. 匿名，作者不详〖前缀 1. an–; 词根 145. onym〗
articulate adj. 发音清晰的，善于表达的；v. 清晰地发音，明确表达
aspire v. 向往，有志于〖词根 200. spir〗
austere adj. 朴素的
berate v. 严厉责备，痛斥〖前缀 10. be–〗
bombastic 夸夸其谈的，空洞的
buzz n. 嗡嗡声，嘈杂声
celebrate v. 庆祝，祝贺，歌颂

collapse v. 坍塌，塌陷

compatible adj. 相容的，兼容的〖前缀 16. com–; 词根 153. pati〗

consensus n. 同意，一致〖前缀 16. con–; 词根 186. sens〗

constituent n. 选民；成分

contradiction n. 反驳；矛盾〖前缀 17. contra–; 词根 55. dict〗

contrary adj. 相反的，反对的，对立的〖前缀 17. contra–〗

counterpose v. 使对立〖前缀 17. counter–; 词根 170. pos〗

deficiency n. 缺陷，不足〖前缀 18. de–; 词根 65. fic〗

dent v. 削弱

discord n. & v. 不和，不一致〖前缀 20. dis–; 词根 41. cord〗

discourse n. 论述，论文

diverse adj. 不同的，多种多样的〖前缀 19. di–; 词根 226. vers〗

diversity n. 多样化，多样性〖前缀 19. di–; 词根 226. vers〗

dynamite n. 炸药；v. 破坏，炸毁

efficacy n. 功效，效力〖前缀 ef (= 26. ex–); 词根 65. fic〗

elusive adj. 逃避的，难以捉摸的

emerge v. 出现，浮现

envisage v. 想象，设想〖前缀 24. en–; 词根 229. vis〗

epistemological adj. 认识论的

equivocal adj. 模棱两可的

extension n. 伸展，扩大，延长〖前缀 26. ex–; 词根 208. tens〗

fabric n. 织物；构造，结构

forge v. 锻制，锻造

genteel adj. 文雅的，有礼貌的，上流社会的

geometrical adj. 几何的，几何学的〖词根 geo, 地；词根 126. metr〗

hybrid n. 杂种，混合物；adj. 杂种的，混合的

ill–founded adj. 无理由的，站不住脚的

illuminate v. 照亮，照明；阐明，说明〖词根 116. lumin〗

imperialism n. 帝国主义，帝制，皇帝的统治

implicit adj. 含蓄的〖前缀 33. im–; 词根 169. pli〗

imply v. 暗示〖前缀 33. im–; 169. pli〗

inchoate adj. 才开始的，未完成的

inconsistent adj. 不一致的，不调和的〖前缀 34. in–〗

incremental adj. 增加的〖前缀 34. in–; 词根 45. cre〗

infancy n. 婴儿期，幼年时代；初期

inform v. 对……有影响；使活跃，使有生气；告诉，通知

inhumane adj. 无人道的，不仁慈的〖前缀 34. in–; 词根 93. hum〗

intervene v. 干涉，介入〖前缀 35. inter–; 词根 222. ven〗

intimidate v. 恐吓，威胁

irrelevant adj. 不相关的，不中肯的；离题的，不重要的〖前缀 36. ir–〗

liberal adj. 自由的，慷慨的〖词根 108. liber〗

liberalism n. 自由主义〖词根 108. liber〗

literally adv. 逐字地，照字面地；确实地，真正地〖词根 110. liter〗

messy adj. 凌乱的，散乱的

mislead v. 把……带错路，使误入歧途，使误解〖前缀 41. mis–〗

normative adj. 标准的，规范的〖词根 137. norm〗

optimistic adj. 乐观的

organic adj. 有机的，自然发展的

orientation n. 方向，定位〖词根 148. ori〗

orthodox adj. 正统的〖词根 149. ortho〗

overarch v. 使成拱形；成为……的中心，支配

paradigmatic adj. 范例的

pared–down adj. 压缩的，简化的

persuasive adj. 有说服力的，劝说的，劝诱的

piecemeal adv. 一件一件地，逐渐地；adj. 一件一件的，逐渐的

plural adj. 复数的；多数的

pluralism n. 多元化，多元性

posit v. 假定，设想

presuppose v. 预先假定〖前缀 57. pre–〗

promote v. 促进，推进，提升〖前缀 57. pre–; 词根 132. mot〗

puritanical adj.（贬）清教徒式的，道德极严格的

pursuit n. 追赶，追逐，追求

reaction n. 反应〖词根 2. act〗

relativism n. 相对论，相对主义

repudiate v. 否认，拒绝接受

rigorous adj. 严厉的，严格的，严酷的，严密的

self–defeating adj. 自我挫败的，违背自己利益的，自我拆台的〖前缀 64. self–〗

skeptical adj. 怀疑的

slum n. 贫民窟，贫民区

soulless adj. 没有生气的，呆板的，乏味的

sterile adj. 贫瘠的；不孕的

tension n. 紧张，不安〖词根 208. tens〗

uncompromising adj. 不妥协的，坚定的〖前缀 74. un-〗

undecidable adj. 不可以决定的〖前缀 74. un-〗

underpin v. 加强……的基础〖前缀 75. under-〗

underpinning n. 基础，基础材料，基础结构〖前缀 75. under-〗

untenable adj. 站不住脚的，难以防守的

utopia n. 乌托邦（理想中美好的社会）

vitality n. 活力，生命力〖词根 230. vit〗

warrant n. 正当理由；v. 使有正当理由，成为……的根据；保证

Complex Sentences

[1] According to some contemporary thinkers, over the last twenty years or so **there has been a sea change** in western thought and culture from "modernism" to "postmodernism"—*a change so significant as to represent something akin to a Kuhnian paradigm shift.*

黑体字部分为句子的主干，in 引导的介词短语与 from 引导的介词短语均作定语，均修饰前面的 a sea change，破折号后为解释性插入语，其中下划线部分为定语，修饰其前的 a change。

[2] Thus according to one of the leading writers about postmodernism, Charles Jencks (1991), **the modern era ended** *when, on 15 July 1972,* ***the Pruitt-Igoe housing estate in St Louis (USA)****—which had earlier won an award as an exemplar of modern architecture and town planning-****was dynamited and destroyed*** *by the local city authority.*

正体黑体字部分为句子的主干，斜体字部分为 when 引导的时间状语从句，斜体黑体字部分为时间状语从句的主干，两个破折号之间为解释性插入语，事实上为 which 引导的定语从句，修饰其前的 the Pruitt-Igoe housing estate in St Louis (USA)。

[3] **It is these values** which allow us to see Venturi (in relation to architecture) and Jacobs (in relation to town planning) as early postmodernists, for they argued for complexity and diversity as opposed to the sterile simplifications of modernism.

黑体字部分为句子的主干，which 引导的从句为定语从句，修饰前面的 these values，括号内容为解释性插入语，for 引导的从句为原因状语从句。

[4] If this is true, then **it suggests** *that* ***one of the most important tasks*** *facing town planning theorists now* ***is the development of better theory*** *about the environmental qualities which town planning practice should help bring into being.*

If 引导的从句为条件状语从句，正体黑体字部分为主语和谓语，斜体字部分为 that 引导的宾语从句，斜体黑体字部分为宾语从句的主干，facing 引导的现在分词短语作定语，修饰其前的

one of the most important tasks，about 引导的介词短语作定语，修饰前面的 better theory，which 引导的从句为定语从句，亦修饰 better theory。

Exercises

[1] 将课文中下划线句子或段落译成中文。

[2] 读熟或背诵上述复杂长句。

[3] 解析下列单词 [解析单词是指根据单词中包含的词根、前缀推导出单词的含义，例如，apathy（n. 漠然，冷淡），解析式为：a-（无）+ path（感情）+ y →无感情→冷漠，冷淡]。解析的单词选自上文"**New Words**"。

anonymity n. 匿名，作者不详	illuminate v. 照亮，照明；阐明，说明
berate v. 严厉责备，痛斥	implicit adj. 含蓄的
consensus n. 同意，一致	mislead v. 把……带错路，使误入歧途，使误解
counterpose v. 使对立	tension n. 紧张，不安
discord n. & v. 不和，不一致	vitality n. 活力，生命力

Unit 16

New Urbanism

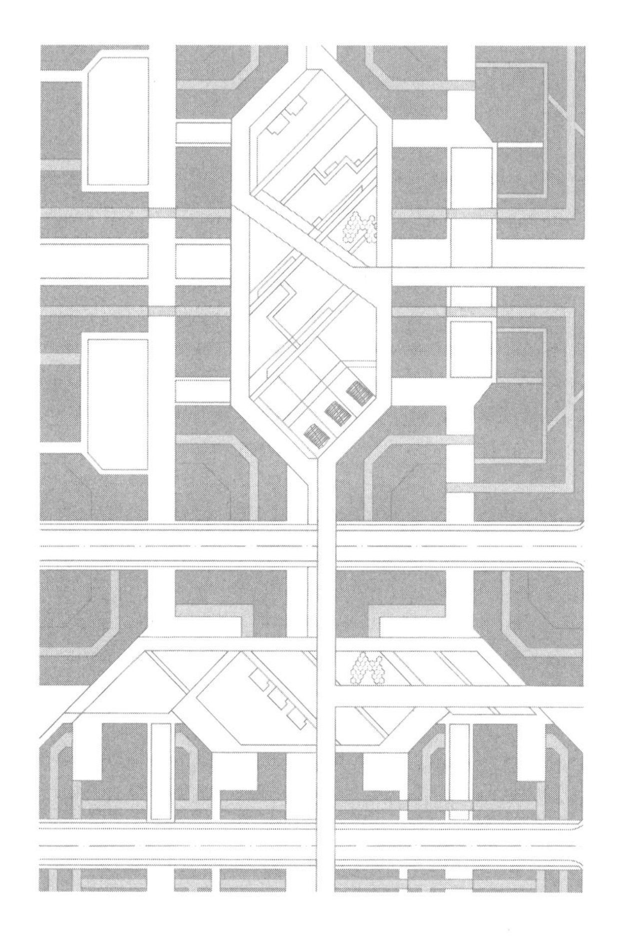

New Urbanism is an urban design movement which **promotes** environmentally friendly habits by creating walkable neighborhoods containing a wide range of housing and job types. It arose in the United States in the early 1980s, and has gradually influenced many aspects of real estate development, urban planning, and **municipal** land–use strategies.

New Urbanism is strongly influenced by urban design practices that were **prominent** until the rise of the automobile prior to World War II; it **encompasses** ten basic principles such as traditional neighborhood design (TND) and **transit–oriented** development (TOD). These ideas can all be circled back to two concepts: building a sense of community and the development of **ecological** practices.

The organizing body for New Urbanism is the Congress for the New Urbanism, founded in 1993. Its foundational text is the Charter of the New Urbanism, which begins:

> We **advocate** the **restructuring** of public policy and development practices to support the following principles: neighborhoods should be **diverse** in use and population; communities should be designed for the **pedestrian** and transit as well as the car; cities and towns should be shaped by physically defined and **universally accessible** public spaces and community **institutions**; urban places should be **framed** by architecture and landscape design that **celebrate** local history, climate, **ecology**, and building practice.

New Urbanists support: regional planning for open space; **context–appropriate** architecture and planning; **adequate provision** of **infrastructure** such as sporting facilities, libraries and community centers; and the balanced development of jobs and housing. They believe their strategies can reduce traffic **congestion** by encouraging the population to ride bikes, walk, or take the train. They also hope that this set up will increase the supply of affordable housing and **rein** in suburban **sprawl**. The Charter of the New Urbanism also covers issues such as historic **preservation**, safe streets, green building, and the re-development of brownfield land. The ten Principles of Intelligent Urbanism also phrase guidelines for new urbanist approaches.

Architecturally, new urbanist developments are often accompanied by New Classical, postmodern, or **vernacular** styles, although that is not always the case.

1 Background

Until the mid 20th century, cities were generally organized into and developed around mixed-use walkable neighborhoods. For most of human history this meant a city that was entirely walkable, although with the development of mass transit the reach of the city **extended** outward along transit lines, allowing for the growth of new pedestrian communities such as streetcar suburbs. But with the **advent** of cheap automobiles and favorable government policies, attention began to shift away from cities and towards ways of growth more focused on the needs of the car. Specifically, after World War II urban planning largely centered around the use of municipal zoning **ordinances** to **segregate** residential from commercial and industrial development, and focused on the construction of low-density single-family **detached** houses as the preferred housing format for the growing middle class. *The physical **separation** of where people live from where they work, shop and frequently spend their recreational time, together with low housing density, which often drastically reduced population density relative to historical norms, made automobiles **indispensable** for practical transportation and contributed to the **emergence** of a culture of automobile dependency.*

This new system of development, with its **rigorous** separation of uses, arose after World War II and became known as "conventional suburban development" or **pejoratively** as urban **sprawl**. The majority of U.S. citizens now live in suburban communities built in the last fifty years, and automobile use per capita has **soared**.

Although New Urbanism as an organized movement would only arise later, a number of **activists** and thinkers soon began to **criticize** the modernist planning techniques being put into practice. Social philosopher and historian Lewis Mumford criticized the "anti-urban" development of post-war America. *The Death and Life of Great American Cities, written by Jane Jacobs in the early 1960s, called for planners to reconsider the single-use housing projects, large car-dependent* ***thoroughfares****, and segregated commercial centers that had become the "norm".*

Rooted in these early **dissenters**, the ideas behind New Urbanism began to **solidify** in the 1970s and 1980s with the urban visions and theoretical models for the reconstruction of the "European" city proposed by architect Leon Krier, and the pattern language theories of Christopher Alexander. The term "new urbanism" itself started being used in this context in the mid-1980s, but it wasn't until the early 1990s that it was commonly written as a proper noun **capitalized**.

In 1991, the Local Government Commission, a private **nonprofit** group in Sacramento, California, invited architects Peter Calthorpe, Michael Corbett, Andrés Duany, Elizabeth Moule, Elizabeth Plater-Zyberk, Stefanos Polyzoides, and Daniel Solomon to develop a set of community principles for land use planning. Named the Ahwahnee Principles (after Yosemite National Park's Ahwahnee Hotel), the commission presented the principles to about one hundred government officials in the fall of 1991, at its first Yosemite Conference for Local Elected Officials.

Calthorpe, Duany, Moule, Plater-Zyberk, Polyzoides, and Solomon founded the Chicago-based Congress for the New Urbanism in 1993. The CNU has grown to more than three thousand members, and is the leading international organization promoting New Urbanist design principles. It holds **annual** Congresses in various U.S. cities.

In 2009, co-founders Elizabeth Moule, Hank Dittmar, and Stefanos Polyzoides authored *the* ***Canons*** *of Sustainable Architecture and Urbanism* to clarify and detail the relationship between New Urbanism and sustainability. The Canons are "a set of operating principles for human settlement that reestablish the relationship between the art of building, the making of community, and the **conservation** of our natural world". They promote the use of **passive** heating and cooling solutions, the use of locally obtained materials, and in general, a "culture of permanence".

New Urbanism is a broad movement that **spans** a number of different disciplines and geographic scales. And while the conventional approach to growth remains dominant, New Urbanist principles have become increasingly influential in the fields of planning, architecture, and public policy.

2 Defining elements

Andrés Duany and Elizabeth Plater-Zyberk, two of the founders of the Congress for the

New Urbanism, observed mixed-use streetscapes with corner shops, front **porches**, and a **diversity** of well-crafted housing while living in one of the Victorian neighborhoods of New Haven, Connecticut. They and their colleagues observed patterns including the following:

· The neighborhood has a **discernible** center. This is often a square or a green and sometimes a busy or memorable street corner. A transit stop would be located at this center.

· Most of the dwellings are within a five-minute walk of the center, an average of roughly 0.25 miles (0.40 km).

· There are a variety of dwelling types—usually houses, rowhouses, and apartments—so that younger and older people, singles and families, the poor and the wealthy may find places to live.

· At the edge of the neighborhood, there are shops and offices of sufficiently varied types to supply the weekly needs of a household.

· A small **ancillary** building or **garage** apartment is permitted within the backyard of each house. It may be used as a rental unit or place to work (for example, an office or craft workshop).

· An elementary school is close enough so that most children can walk from their home.

· There are small playgrounds **accessible** to every dwelling—not more than a tenth of a mile away.

· Streets within the neighborhood form a connected network, which **disperses** traffic by providing a variety of pedestrian and vehicular routes to any destination.

· The streets are relatively narrow and shaded by rows of trees. This slows traffic, creating an environment suitable for pedestrians and bicycles.

· Buildings in the neighborhood center are placed close to the street, creating a well-defined outdoor room.

· Parking lots and garage doors rarely front the street. Parking is relegated to the rear of buildings, usually **accessed** by **alleys**.

· Certain prominent sites at the **termination** of street **vistas** or in the neighborhood center are **reserved** for civic buildings. These provide sites for community meetings, education, and religious or cultural activities.

3 Examples

3.1 United States

New Urbanism is having a growing influence on how and where **metropolitan** regions choose to grow. At least fourteen large-scale planning **initiatives** are based on the principles of linking transportation and land-use policies, and using the neighborhood as the fundamental building block of a region. Miami, Florida, has adopted the most **ambitious** New Urbanist-based zoning code reform yet undertaken by a major U.S. city.

More than six hundred new towns, villages, and neighborhoods, following New Urbanist principles, have been planned or are currently under construction in the U.S. Hundreds of new, small-scale, urban and suburban infill projects are under way to reestablish walkable streets and blocks. In Maryland and several other states, New Urbanist principles are an **integral** part of smart growth legislation.

In the mid-1990s, the U.S. Department of Housing and Urban Development (HUD) adopted the principles of the New Urbanism in its **multibillion**-dollar program to rebuild public housing projects nationwide. New Urbanists have planned and developed hundreds of projects in infill locations. Most were driven by the private sector, but many, including HUD projects, used public money.

3.2 Other countries

New Urbanism is closely related to the urban village movement in Europe. They both occurred at similar times and share many of the same principles although urban villages has an emphasis on traditional city planning. In Europe many brown-field sites have been redeveloped since the 1980s following the models of the traditional city neighborhoods rather than Modernist models. One well-publicized example is Poundbury in England, a suburban **extension** to the town of Dorchester, which was built on land owned by the Duchy of Cornwall under the overview of Prince Charles. The **original** masterplan was designed by Leon Krier. A report carried out after the first phase of construction found a high degree of satisfaction by residents, although the **aspirations** to reduce car dependency had not been successful. Rising house prices and a perceived **premium** have made the open market housing **unaffordable** for many local people.

The Council for European Urbanism (CEU), formed in 2003, shares many of the same aims as the U.S.'s New Urbanists. CEU's Charter is a development of the Congress for the New Urbanism Charter **revised** and reorganized to relate better to European conditions. An Australian organization, Australian Council for New Urbanism has since 2001 run conferences and events to promote New Urbanism in that country. A New Zealand Urban Design **Protocol** was created by the Ministry for the Environment in 2005.

There are many developments around the world that follow New Urbanist principles to a greater or lesser extent.

Notes

[1] 本文源自维基百科（From Wikipedia，the free encyclopedia，https://www.wikipedia.org/）。

[2] 本文主要相关文献：彼得· 卡尔索普, 等, 2019; 马交国, 等, 2018; 张衔春, 等, 2016; 张雅鹏, 等, 2017; Calthorpe，et al，2001; Duany，et al，2000; Dutton，2000; Katz，1994。

[3] 课文中下划线部分为有关城市规划理论的重要论述，需强化阅读，每篇课文习题部分均要求将下划线句子或段落译成中文。

[4] 课文中下划线斜体字部分为复杂长句，其解析详见下文“**Complex Sentences**”。

New Words

access n. 接近，进入，通路；v. 接近，使用〖前缀 3. ac–; 词根 30. cess〗

accessible adj. 易达到的，易受影响的〖前缀 3. ac–; 词根 30. cess〗

activist n. 激进主义分子，积极分子〖词根 2. act〗

adequate adj. 足够的；适当的

advent n. 出现，到来〖前缀 3. ad–; 词根 222. vent〗

advocate v. 拥护，支持，鼓吹；n. 支持者，拥护者〖前缀 3. ad–; 词根 232. voc〗

alley n. 胡同，小巷，小径

ambitious adj. 有雄心的，有野心的〖词根 5. ambi–〗

ancillary adj. 辅助的，补充的，附属的

annual adj. 每年的，一年生的〖词根 11. ann〗

appropriate adj. 适当的，恰当的

aspiration n. 强烈的愿望〖词根 200. spir〗

canon n. 标准，准则，原则；教规，宗教法规

capitalize v. 用大写字母写或印刷；使……资本化〖词根 25. cap〗

celebrate v. 庆祝，祝贺，歌颂

congestion n. 拥挤，阻塞

conservation n. 保存，保护〖前缀 16. con–; 词根 188. serv〗

context n. 上下文，背景，语境，环境

criticize v. 批评，评论〖词根 48. crit〗

detached adj. 分离的，分开的〖前缀 18. de–; 词根 204. tach〗

discernible adj. 可识别的，可辨别的

disperse v. 分散，散开，散播〖前缀 20. dis–〗

dissenter n. 持异议者，持不同政见者〖前缀 20. dis–; 词根 186. sent〗

diverse adj. 不同的，多种多样的〖前缀 19. di–; 词根 226. vers〗

diversity n. 多样化，多样性〖前缀 19. di–; 词根 226. vers〗

ecological adj. 生态的，生态学的〖词根 57. eco〗

ecology n. 生态学〖词根 57. eco〗

emergence n. 出现，发生

encompass v. 包含，包括；围绕，包围〖前缀 24. en–〗

extend v. 延伸，扩大，推广〖前缀 26. ex–; 词根 208. tend〗

extension n. 伸展，扩大，延长〖前缀 26. ex–; 词根 208. tens〗

frame v. 构造，建造，塑造；制定，拟出，设计

garage n. 车库

indispensable adj. 不可缺少的，绝对必要的〖前缀 34. in–〗

infrastructure n. 基础设施，基础建设〖词根 201. struct〗

initiative n. 主动，首创精神

institution n. 制度，机制；社会公共机构，学校

integral adj. 构成整体所必需的，完整的

metropolitan adj. 大都会的，大城市的

multibillion adj. 数十亿的〖前缀 43. multi–〗

municipal adj. 市的，市政的〖词根 134. mun〗

nonprofit adj. 非营利的〖前缀 45. non–〗

ordinance n. 条例，法令

orient v. 标定方向，使……向东方；adj. 东方的，新生的；n. 东方〖词根 148. ori〗

original adj. 最初的，原始的；有创意的〖词根 148. ori〗

passive adj. 被动的，消极的

pedestrian n. 步行者，行人；adj. 徒步的，平淡无奇的〖词根 156. ped〗

pejoratively adv. 轻蔑地

porch n. 门廊，游廊，走廊

premium adj. 高级的，优质的 / at a premium 非常珍贵的，奇缺的

preservation n. 保存，保留，保护〖词根 188. serv〗

prominent adj. 杰出的，著名的

promote v. 促进，推进，提升〖前缀 58. pro–; 词根 132. mot〗

protocol n.（条约等的）草案，草约；议定书，协议

provision n. 规定，条款；供应

rein n. 缰绳，驾驭，控制

reserve n. 储备，保留〖词根 188. serv〗

restructure v. 改组，重建〖前缀 61. re–; 词根 201. struct〗

revise v. 修订，修正〖前缀 61. re–; 词根 229. vis〗

rigorous adj. 严格的，严密的

segregate v. 分开，分离〖前缀 63. se-; 词根 86. greg〗

separation n. 分离，隔离〖前缀 63. se-〗

soar v. 高飞，猛增

solidify v. 凝固，固化；团结〖词根 195. solid〗

span v. 横跨，跨越

sprawl n. 蔓延，随意扩展

termination n. 终止，结束〖词根 209. term〗

thoroughfare n. 大道，大街

transit n. 通过，经过；运输，运输路线，公共交通系统〖前缀 71. trans-〗

unaffordable adj. 买不起的，负担不起的〖前缀 74. un-〗

universally adv. 普遍地，一般地〖前缀 76. uni-〗

vernacular adj. 本国的，本地的，乡土的

vista n. 远景，街景〖词根 229. vis〗

Complex Sentences

[1] **The physical separation of where people live from where they work, shop and frequently spend their recreational time**, together with low housing density, which often drastically reduced population density relative to historical norms, ***made automobiles indispensable for practical transportation*** **and** ***contributed to the emergence of a culture of automobile dependency***.

正体黑体字部分为句子的主语，主语很长。斜体黑体字部分为句子的谓语和宾语，本句为双谓语双宾语结构。together 引导的介词短语作定语，修饰前面的 the physical separation，which 引导的定语从句修饰前面的 low housing density。

[2] **The Death and Life of Great American Cities**, written by Jane Jacobs in the early 1960s, **called for planners to reconsider the single-use housing projects, large car-dependent thoroughfares, and segregated commercial centers that had become the "norm"**.

黑体字部分为句子的主干（主语、谓语和宾语），其中 The Death and Life of Great American Cities 为著作名称，作主语，written 引导的过去分词短语作定语，修饰主语，that 引导的从句为定语从句，修饰其前的三个名词性短语。

Exercises

[1] 将课文中下划线句子或段落译成中文。

[2] 读熟或背诵上述复杂长句。

[3] 解析下列单词[解析单词是指根据单词中包含的词根、前缀推导出单词的含义，例如，apathy（n. 漠然，冷淡），解析式为：a-（无）+ path（感情）+ y →无感情→冷漠，冷淡]。解析的单词选自上文“**New Words**”。

annual adj. 每年的，一年生的	infrastructure n. 基础设施，基础建设
criticize v. 批评，评论	multibillion adj. 数十亿的
detached adj. 分离的，分开的	restructure v. 改组，重建
disperse v. 分散，散开，散播	revise v. 修订，修正
dissenter n. 持异议者，持不同政见者	termination n. 终止，结束
indispensable adj. 不可缺少的，绝对必要的	vista n. 远景，街景

Unit 17

Urban Sociology

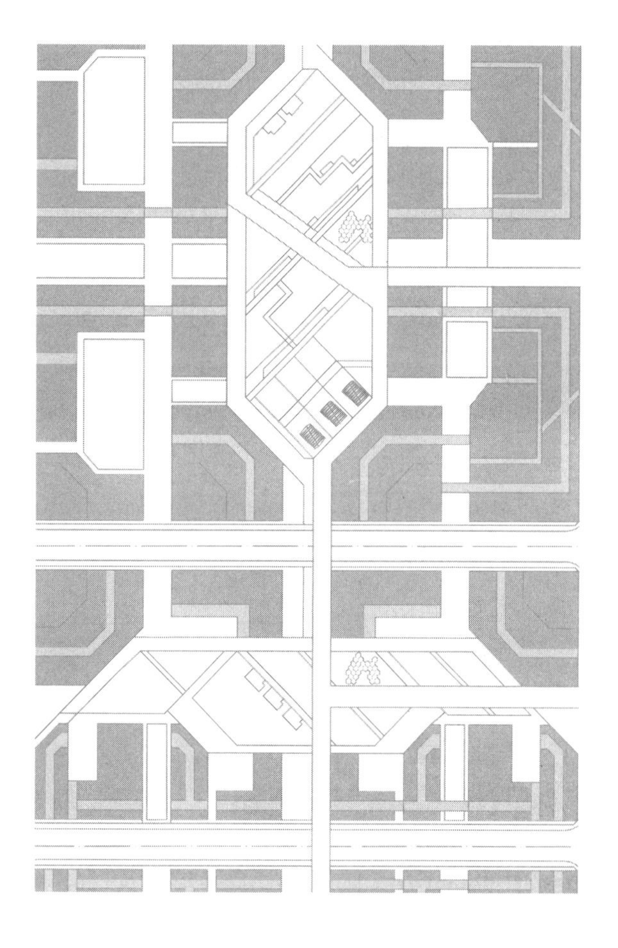

Urban sociology is the sociological study of life and human interaction in **metropolitan** areas. It is a **normative** discipline of sociology seeking to study the structures, environmental processes, changes and problems of an urban area and by doing so provide inputs for urban planning and policy making. In other words, it is the sociological study of cities and their role in the development of society. Like most areas of sociology, urban sociologists use statistical analysis, observation, social theory, interviews, and other methods to study a range of topics, including **migration** and **demographic** trends, economics, poverty, race relations and economic trends.

The philosophical foundations of modern urban sociology ***originate*** *from the work of sociologists such as Karl Marx, Ferdinand Tönnies, Émile Durkheim, Max Weber and Georg Simmel who studied and theorized the economic, social and cultural processes of urbanization and its effects on social* ***alienation****, class formation, and the production or* ***destruction*** *of collective and individual identities.*

These theoretical foundations were further expanded upon and analyzed by a group of sociologists and researchers who worked at the University of Chicago in the early twentieth century. *In what became known as the Chicago School of sociology the work of Robert Park, Louis Wirth and Ernest Burgess on the* ***inner*** *city of Chicago revolutionized the purpose of urban research in sociology but also the development of human geography through its use of quantitative and* ***ethnographic*** *research methods.* The importance of the theories developed by the Chicago School within urban sociology have been **critically sustained** and **critiqued** but still remain one of the most significant historical advancements in understanding urbanization and the city within the social sciences.

1 Development and rise of urban sociology

Urban sociology rose to prominence within the academy in North America through a group of sociologists and theorists at the University of Chicago from 1915 to 1940 in what became known as the Chicago School of Sociology. The Chicago School of Sociology combined sociological and **anthropological** theory with ethnographic fieldwork in order to understand how individuals interact within urban social systems. Unlike the primarily macro–based sociology that had marked earlier subfields, members of the Chicago School placed greater emphasis on micro–scale social interactions that sought to provide **subjective** meaning to how humans interact under structural, cultural and social conditions. The theory of **symbolic** interaction, the basis through which many methodologically **groundbreaking ethnographies** were **framed** in this period, took primitive shape alongside urban sociology and shaped its early methodological leanings. Symbolic interaction was **forged** out of the writings of early micro–sociologists George Mead and Max Weber, and sought to frame how individuals interpret symbols in everyday interactions. With early urban sociologists framing the city as a "superorganism", the concept of symbolic interaction aided in **parsing** out how individual communities contribute to the **seamless** functioning of the city itself.

Scholars of the Chicago School originally sought to answer a single question: how did an increase in urbanism during the time of the Industrial Revolution contribute to the magnification of contemporary social problems? Sociologists centered on Chicago due to its **tabula rasa** state, having expanded from a small town of 10,000 in 1860 to an urban **metropolis** of over two million in the next half–century. Along with this expansion came many of the era's emerging social problems–ranging from issues with concentrated homelessness and harsh living conditions to the low wages and long hours that characterized the work of the many newly arrived European **immigrants**. Furthermore, unlike many other metropolitan areas, Chicago did not expand outward at the edges as predicted by early expansionist theorists, but instead **reformatted** the space available in a **concentric** ring pattern. As with many modern cities the business district occupied the city center and was surrounded by **slum** and **blighted** neighborhoods, which were further surrounded by workingmens' homes and the early forms of the modern suburbs. Urban theorists suggested that these spatially distinct regions helped to **solidify** and **isolate** class relations within the modern city, moving the middle class away from the urban core and into the privatized environment of the outer suburbs.

Due to the high **concentration** of first-generation immigrant families in the inner city of Chicago during the early 20th century, many prominent early studies in urban sociology focused upon the **transmission** of immigrants' native culture roles and norms into new and developing environments. Political participation and the rise in inter-community organizations were also frequently covered in this period, with many metropolitan areas adopting census techniques that allowed for information to be stored and easily accessed by participating institutions such as the University of Chicago. Park, Burgess and McKenzie, professors at the University of Chicago and three of the earliest proponents of urban sociology, developed the Subculture Theories, which helped to explain the often-positive role of local institutions on the formation of community acceptance and social ties. When race relations break down and expansion **renders** one's community members **anonymous**, as was proposed to be occurring in this period, the inner city becomes marked by high levels of social disorganization that prevent local ties from being established and maintained in local political arenas.

The rise of urban sociology **coincided** with the expansion of statistical **inference** in the behavioural sciences, which helped **ease** its **transition** and acceptance in educational institutions along with other **burgeoning** social sciences. Micro-sociology courses at the University of Chicago were among the earliest and most prominent courses on urban sociological research in the United States.

2 Evolution of urban sociology

The evolution and transition of sociological theory from the Chicago School began to emerge in the 1970s with the publication of Claude Fischer's (1975) *Toward a Theory of Subculture Urbanism* which incorporated Bourdieu's theories on social capital and symbolic capital within the invasion and succession framework of the Chicago School in explaining how cultural groups form, expand and solidify a neighborhood. The theme of transition by subcultures and groups within the city was further expanded by Barry Wellman's (1979) *The Community Question: The Intimate Networks of East Yorkers* which determined the function and position of the individual, institution and community in the urban landscape in relation to their community. Wellman's categorization and incorporation of community focused theories as "Community Lost", "Community Saved", and "Community Liberated" which center around the structure of the urban community in shaping interactions between individuals and facilitating active participation in the local community are explained in detail below:

Community lost: The earliest of the three theories, this concept was developed in the late 19th century to account for the rapid development of industrial patterns that seemingly caused **rifts** between the individual and their local community. Urbanites were claimed to hold networks that were "**impersonal**, **transitory** and **segmental**", maintaining ties in **multiple** social networks while at the same time lacking the strong ties that bound them to any specific group. This disorganization in turn caused members of urban communities to subsist almost solely on secondary **affiliations** with others, and rarely allowed them to rely on other members of the community for assistance with their needs.

Community saved: A critical response to the community lost theory that developed during the 1960s, the community saved argument suggests that **multistranded** ties often emerge in **sparsely-knit** communities as time goes on, and that urban communities often possess these strong ties, **albeit** in different forms. Especially among low-income communities, individuals have a tendency to adapt to their environment and pool resources in order to protect themselves collectively against structural changes. Over time urban communities have tendencies to become "urban villages", where individuals possess strong ties with only a few individuals that connect them to an intricate web of other urbanities within the same local environment.

Community liberated: A cross-section of the community lost and community saved arguments, the community liberated theory suggests that the separation of workplace, residence and **familial kinship** groups has caused urbanites to maintain weak ties in multiple community groups that are further weakened by high rates of residential **mobility**. However, the concentrated number of environments present in the city for interaction increase the likelihood of individuals developing secondary ties, even if they simultaneously maintain distance from **tightly-knit** communities. Primary ties that offer the individual assistance in everyday life form out of sparsely-knit and spatially **dispersed** interactions, with the individual's access to resources dependent on the quality of the ties they maintain within their community.

Along with the development of these theories, urban sociologists have increasingly begun to study the differences between the urban, rural and suburban environment within the last half-century. Consistent with the community liberated argument, researchers have in large part found that urban residents tend to maintain more spatially-dispersed networks of ties than rural or suburban residents. Among lower-income urban residents, the lack of mobility and **communal** space within the city often **disrupts** the formation of social ties and lends itself to creating an **unintegrated** and distant community space. While the high density of networks within the city weakens relations between individuals, it increases the likelihood that at least one individual within a network can provide the primary support found among smaller and more tightly-knit networks. Since the 1970s, research into social networks has focused primarily on the types of ties developed within residential environments. Bonding ties, common of tightly-knit neighborhoods, consist of connections that provide an individual with primary support, such as access to income or upward mobility among a neighborhood organization. Bridging ties, in contrast, are the ties that weakly connect strong networks of individuals together. A group of communities concerned about the placement of a nearby highway may only be connected through a few individuals that represent their views at a community board meeting, for instance.

However, as theory surrounding social networks has developed, sociologists such as Alejandro Portes and the Wisconsin model of sociological research began placing increased **leverage** on the importance of these weak ties. While strong ties are necessary for providing residents with primary services and a sense of community, weak ties bring together elements of different cultural and economic landscapes in solving problems affecting a great number of individuals. As theorist Eric Oliver notes, neighborhoods with vast social networks are also those that most commonly rely on **heterogeneous** support in problem solving, and are also the most politically active.

As the suburban landscape developed during the 20th century and the outer city became a refuge for the wealthy and, later, the burgeoning middle class, sociologists and urban geographers such as Harvey Molotch, David Harvey and Neil Smith began to study the structure and ***revitalization*** *of the most* ***impoverished*** *areas of the inner city.* In their research, impoverished neighborhoods, which often rely on tightly-knit local ties for economic and social support, were found to be targeted by developers for **gentrification** which **displaced** residents living within these communities. Political experimentation in providing these residents with semi-permanent housing and structural support have in many cases eased the transition of low-income residents into stable housing and employment. Yet research covering the social impact of forced movement among these residents has noted the difficulties individuals often have with maintaining a level of economic comfort, which is spurred by rising land values and inter-urban competition between cities. The interaction between inner-city dwellers and middle class **passersby** in such settings has also been a topic of study for urban sociologists.

3 Criticism

Many theories in urban sociology have been criticized, most prominently directed toward the **ethnocentric** approaches taken by many early theorists that lay groundwork for urban studies throughout the 20th century. Early theories that sought to frame the city as an adaptable "superorganism" often disregarded the intricate roles of social ties within local communities, suggesting that the urban environment itself rather than the individuals living within it controlled the spread and shape of the city. For impoverished inner-city residents, the role of highway planning policies and other government-spurred initiatives instituted by the planner Robert Moses and others have been criticized as **unsightly** and **unresponsive** to residential needs.

Some modern social theorists have also been critical toward the apparent shortsightedness that urban sociologists have shown toward the role of culture in the inner city. William Julius Wilson has criticized theory developed throughout the middle of the twentieth century as relying primarily on structural roles of institutions, and not how culture itself affects common aspects of inner-city life such as poverty. The distance shown toward this topic, he argues, presents an incomplete picture of inner-city life. The urban sociological theory is viewed as one important aspect of sociology.

Notes

[1] 本文源自维基百科（From Wikipedia，the free encyclopedia，https://www.wikipedia.org/）。

[2] 本文主要相关文献：顾朝林，等，2013; LeGates，et al，2007; Lin，et al，2004; Macionis，et al，2006; Savage，et al，2003。

[3] 课文中下划线部分为有关城市规划理论的重要论述，需强化阅读，每篇课文习题部分均要求将下划线句子或段落译成中文。

[4] 课文中下划线斜体字部分为复杂长句，其解析详见下文"**Complex Sentences**"。

New Words

affiliation n. 加入，附属；关系

albeit conj. 虽然，即使

alienation n. 离间，疏远〖词根 6. alien〗

anonymous adj. 无名的，匿名的〖前缀 1. an-; 词根 145. onym〗

anthropological 人类学的〖词根 12. anthrop〗

blight v. 凋萎，颓丧

burgeon v. 迅速发展

coincide v. 巧合，重合；一致，相符〖前缀 16. co-; 词根 23. cid〗

communal adj. 公民的，公共的

concentration n. 集中，专心〖前缀 16. con-; 词根 32. centr〗

concentric adj. 同一中心的〖前缀 16. con-; 词根 32. centr〗

critically adv. 批判性地，苛求地〖词根 48. crit〗

critique v. 批判，发表评论〖词根 48. crit〗

demographic adj. 人口统计学的，人口统计的〖词根 54. dem; 词根 84. graph(y)〗

destruction n. 破坏，毁灭，消灭〖前缀 18. de-; 词根 201. struct〗

disperse v. 分散，散开，散播〖前缀 20. dis-〗

displace v. 移动，移走；替换，取代〖前缀 20. dis-〗

disrupt v. 分裂，瓦解；混乱〖前缀 20. dis-; 词根 178. rupt〗

ease v. 缓解，减少

ethnocentric adj. 种族优越感的〖词根 60. ethno; 词根 32. centr〗

ethnographic adj. 人种志的〖词根 60. ethno; 84. graph(y)〗

ethnography n. 人种论，民族志〖词根 60. ethno; 84. graph(y)〗

familial adj. 家族的，家庭的

forge v. 锻制，锻造

frame v. 设计，表达

gentrification n. 中产阶级化

groundbreaking adj. 开创性的，突破性的

heterogeneous adj. 异类的，不同的〖前缀 29. hetero-; 词根 81. gene〗

immigrant n. 移民〖前缀 33. im-; 词根 127. migr〗

impersonal adj. 没有人情味的〖前缀 33. im-〗

impoverished adj. 贫穷的

inference n. 推理，推论

inner adj. 内部的，内心的

isolate v. 隔离，孤立〖词根 194. sol〗

kinship n. 亲属关系

leverage n. 优势，力量；杠杆作用

metropolitan adj. 大城市的，大都会的

metropolis n. 大都市，大都会

migration n. 迁移，移居〖词根 127. migr〗

mobility n. 流动性，移动性〖词根 130. mob〗

multiple adj. 多重的，复杂的〖前缀 43. multi-〗

multistranded adj.（绳子）多股的〖前缀 43. multi-〗

normative adj. 标准的，规范的

originate v. 发源，发起；创始，发明〖词根 148. ori〗

parse v. 评论性分析

passersby n. 过路人，行人

reformat v. 重定格式〖前缀 61. re–; 词根 76. form〗

render v. 使得，使变为

revitalization n. 新生，复兴〖前缀 61. re–; 词根 230. vit〗

rift n. 裂缝，裂口，不和

seamless adj. 无缝的，无漏洞的

segmental adj. 部分的〖前缀 63. se–〗

slum n. 贫民窟，贫民区

solidify v. 凝固，固化；团结〖词根 195. solid〗

sparsely–knit adj. 联系松散的

subjective adj. 主观的

sustain v. 维持，支撑

symbolic adj. 象征的，象征性的

tabula rasa n. 纯洁质朴状态；光板

tightly–knit adj. 联系紧密的

transition n. 过渡，转变，变迁〖前缀 71. trans–〗

transitory adj. 短暂的〖前缀 71. trans–〗

transmission n. 传送，传播〖前缀 71. trans–; 词根 129. miss〗

unresponsive adj. 无答复的，反应迟钝的〖前缀 74. un–〗

unsightly adj. 难看的，不好看的〖前缀 74. un–〗

Complex Sentences

[1] **The philosophical foundations of modern urban sociology originate from the work of sociologists** such as Karl Marx, Ferdinand Tönnies, Émile Durkheim, Max Weber and Georg Simmel **who** studied and theorized the economic, social and cultural processes of urbanization and its effects on social alienation, class formation, and the production or destruction of collective and individual identities.

黑体字部分为句子的主干，such as 引导的介词短语作定语，who 引导的从句为定语从句，均修饰前面的 sociologists。

[2] In what became known as the Chicago School of sociology **the work of Robert Park, Louis Wirth and Ernest Burgess on the inner city of Chicago revolutionized the purpose of urban research in sociology but also the development of human geography** through its use of quantitative and ethnographic research methods.

黑体字部分为句子的主干，the work of Robert Park，Louis Wirth and Ernest Burgess on the inner city of Chicago 为主语，revolutionized 为谓语，the purpose of urban research in sociology but also the development of human geography 为宾语。

[3] **As** the suburban landscape developed during the 20th century and the outer city became a refuge for the wealthy and, later, the burgeoning middle class, **sociologists and urban geographers** such as Harvey Molotch, David Harvey and Neil Smith **began to study the structure and revitalization of the most impoverished areas of the inner city.**

As 引导的从句为状语从句，黑体字部分为句子主干，其中 sociologists and urban geographers 为主语，such as Harvey Molotch，David Harvey and Neil Smith 作定语，修饰主语，began 为谓语，动词不定式短语 to study the structure and revitalization of the most impoverished areas of the inner city 为宾语。

Exercises

[1] 将课文中下划线句子或段落译成中文。

[2] 读熟或背诵上述复杂长句。

[3] 解析下列单词 [解析单词是指根据单词中包含的词根、前缀推导出单词的含义，例如，apathy（n. 漠然，冷淡），解析式为：a-（无）+ path（感情）+ y →无感情→冷漠，冷淡]。解析的单词选自上文“**New Words**”。

coincide v. 巧合，重合；一致，相符	ethnographic adj. 人种志的
concentration n. 集中，专心	heterogeneous adj. 异类的，不同的
concentric adj. 同一中心的	isolate v. 隔离，孤立
demographic adj. 人口统计（学）的	revitalization n. 新生，复兴
destruction n. 破坏，毁灭，消灭	transition n. 过渡，转变，变迁
disrupt v. 分裂，瓦解；混乱	transitory adj. 短暂的
ethnocentric adj. 种族优越感的	transmission n. 传送，传播

Unit 18

Urban Economics

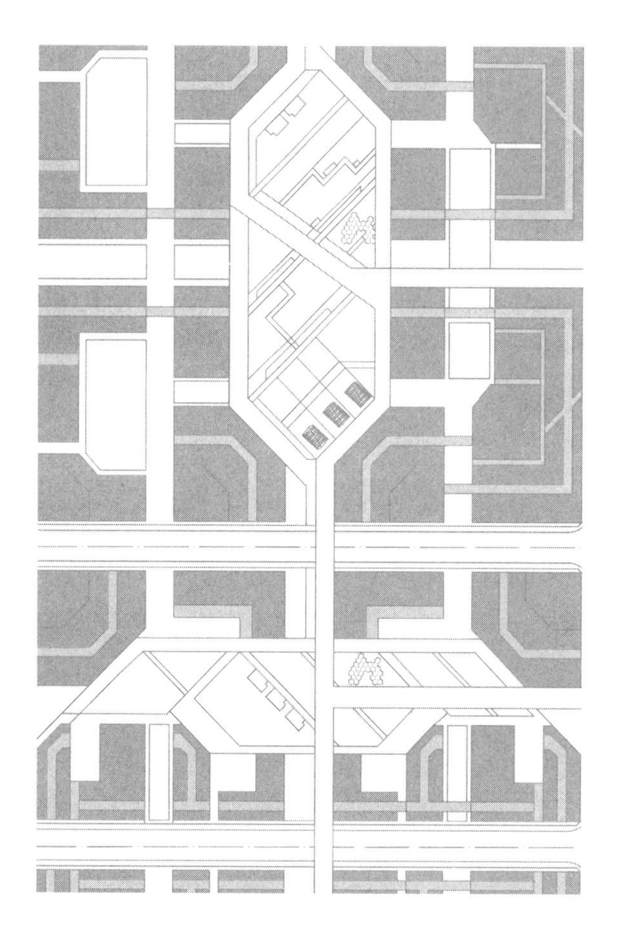

Urban **economics** is broadly the **economic** study of urban areas; as such, it involves using the tools of economics to analyze urban issues such as crime, education, public transit, housing, and local government finance. More narrowly, it is a branch of **microeconomics** that studies urban spatial structure and the location of households and firms.

Much urban economic analysis relies on a particular model of urban spatial structure, the **monocentric** city model pioneered in the 1960s by William Alonso, Richard Muth, and Edwin Mills. While most other forms of **neoclassical** economics do not account for spatial relationships between individuals and organizations, urban economics focuses on these spatial relationships to understand the economic motivations underlying the formation, functioning, and development of cities.

Since its formulation in 1964, Alonso's monocentric city model of a disc–shaped Central Business District (CBD) and surrounding residential region has served as a starting point for urban economic analysis. ***Monocentricity*** *has weakened over time because of changes in technology, particularly, faster and cheaper transportation (which makes it possible for* ***commuters*** *to live farther from their jobs in the CBD) and communications (which allow back–office operations to move out of the CBD).*

Additionally, recent research has sought to explain the **polycentricity** described in Joel Garreau's *Edge City*. Several explanations for polycentric expansion have been proposed and summarized in models that account for factors such as utility gains from lower average land rents and increasing (or constant returns) due to economies of

agglomeration.

1 Introduction

Urban economics is rooted in the location theories of von Thünen, Alonso, Christaller, and Lösch that began the process of spatial economic analysis. Economics is the study of the allocation of scarce resources, and as all economic phenomena take place within a geographical space, urban economics focuses on the allocation of resources across space in relation to urban areas. *Other branches of economics* ***ignore*** *the spatial aspects of decision making but urban economics focuses not only on the location decisions of firms, but also of cities themselves as cities themselves represent centers of economic activity.*

Many spatial economic topics can be analyzed within either an urban or regional economics framework as some economic phenomena primarily affect localized urban areas while others are felt over much larger regional areas. Arthur O'Sullivan believes urban economics is divided into six related themes: market forces in the development of cities, land use within cities, urban transportation, urban problems and public policy, housing and public policy, and local government expenditures and taxes.

2 Market forces in the development of cities

Market forces in the development of cities relate to how the location decision of firms and households causes the development of cities. The nature and behavior of markets depends somewhat on their locations therefore market performance partly depends on geography. If a firm locates in a geographically **isolated** region, their market performance will be different than a firm located in a **concentrated** region. The location decisions of both firms and households create cities that differ in size and economic structure. When industries **cluster**, like in the *Silicon Valley* in *California*, they create urban areas with dominant firms and distinct economies.

By looking at location decisions of firms and households, the urban **economist** is able to address why cities develop where they do, why some cities are large and others small, what causes economic growth and decline, and how local governments affect urban growth. Because urban economics is concerned with asking questions about the nature and workings of the **economy** of a city, models and techniques developed within the field are primarily designed to analyze phenomena that are confined within the limits of a single city.

3 Land use

Looking at land use within **metropolitan** areas, the urban economist seeks to analyze the spatial organization of activities within cities. In attempts to explain observed patterns of land use, the urban economist examines the **intra-city** location choices of firms and households. Considering the spatial organization of activities within cities, urban economics addresses questions in terms of what determines the price of land and why those prices vary across space, the economic forces that caused the spread of employment from the central core of cities **outward**, identifying land-use controls, such as zoning, and interpreting how such controls affect the urban economy.

4 Economic policy

Economic policy is often **implemented** at the urban level thus economic policy is often tied to urban policy. Urban problems and public policy tie into urban economics as the theme relates urban problems, such as poverty or crime, to economics by seeking to answer questions with economic guidance. For example, does the tendency for the poor to live close to one another make them even poorer?

5 Transportation and economics

Urban transportation is a theme of urban economics because it affects land-use patterns as transportation affects the relative **accessibility** of different sites. Issues that tie urban transportation to urban economics include the **deficit** that most **transit** authorities have, and **efficiency** questions about proposed transportation developments such as light-rail. **Megaprojects** such as this have been shown to be **synonymous** with unexpected costs and questionable benefits.

6 Housing and public policy

Housing and public policy relate to urban economics as housing is a **unique** type of **commodity**. Because housing is **immobile**, when a household chooses a dwelling, it is also choosing a location. Urban economists analyze the location choices of households in **conjunction** with the market effects of housing policies. In analyzing housing policies, we make use of market structures e.g., perfect market structure. There are however problems

encountered in making this analysis such as funding, uncertainty, space, etc.

7 Government expenditures and taxes

The final theme of local government **expenditures** and taxes relates to urban economics as it analyzes the efficiency of the **fragmented** local governments **presiding** in metropolitan areas.

Notes

[1] 本文源自维基百科（From Wikipedia, the free encyclopedia, https://www.wikipedia.org/）。

[2] 本文主要相关文献：栾峰，2012; LeGates, et al, 2007; O'Sullivan, 2002。

[3] 课文中下划线部分为有关城市规划理论的重要论述，需强化阅读，每篇课文习题部分均要求将下划线句子或段落译成中文。

[4] 课文中下划线斜体字部分为复杂长句，其解析详见下文"**Complex Sentences**"。

New Words

accessibility n. 可达性〖前缀 3. ac–; 词根 30. cess〗

agglomeration n. 成团，结块

cluster v. 丛生，群聚

commodity n. 商品，日用品

commuter n. 乘公交车辆上下班者，经常乘车往返者〖前缀 16. com–; 词根 135. mut〗

concentrate v. 集中，聚集，专心于〖前缀 16. con–; 词根 32. centr〗

conjunction n. 连接，联合〖前缀 16. con–; 词根 101. junct〗

deficit n. 赤字，亏损〖前缀 18. de–; 词根 65. fic〗

economic adj. 经济学的，经济（上）的〖词根 57. eco〗

economics n. 经济学〖词根 57. eco〗

economist n. 经济学家，经济学研究者〖词根 57. eco〗

economy n. 经济，经济制度〖词根 57. eco〗

efficiency n. 效率，效能〖前缀 ef–(= 26. ex–); 词根 65. fic〗

encounter v. 遭遇，不期而遇〖前缀 24. en–; 前缀 17. counter–〗

expenditure n. 花费，支出〖词根 159. pend〗

fragmented adj. 成碎片的，片段的〖词根 77. frag〗

ignore v. 忽视，不顾〖词根 gnor〗
immobile adj. 固定的，不变的，稳定的〖前缀 33. im–; 词根 130. mob〗
implement v. 实施，执行
intra–city adj. 市内的〖前缀 intra–, 在内，内部〗
isolate v. 隔离，孤立〖194. sol〗
macroeconomics n. 宏观经济学〖前缀 37. macro–; 词根 57. eco〗
megaproject n. 特大工程，特大项目〖前缀 mega–, 大，强〗
metropolitan adj. 大城市的，大都会的
microeconomics n. 微观经济学〖前缀 40. micro–; 词根 57. eco〗
monocentric adj. 单一中心的〖前缀 42. mono–; 词根 32. centr〗
monocentricity n. 单一中心，单一中心性〖前缀 42. mono–; 词根 32. centr〗
neoclassical adj. 新古典主义的〖前缀 44. neo–〗
outward adv. 向外，在外〖前缀 47. out–〗
polycentricity n. 多中心，多中心性〖前缀 55. poly–; 词根 32. centr〗
preside v. 主持，指挥〖前缀 57. pre–; 词根 189. sid〗
synonymous adj. 同义的〖前缀 70. syn–; 词根 145. onym〗
transit n. 通过，经过；运输，运输路线，公共交通系统〖前缀 71. trans–〗
unique adj. 唯一的，独一无二的〖前缀 76. uni–〗

Complex Sentences

[1] **Monocentricity has weakened** over time because of changes in technology, particularly, faster and cheaper transportation (which makes it possible for commuters to live farther from their jobs in the CBD) and communications (which allow back–office operations to move out of the CBD).

黑体字部分为句子的主干（主语和谓语），faster and cheaper transportation and communications 为 changes in technology 的同位语，括号内容为解释性插入语，亦即定语从句。

[2] Other branches of economics ignore the spatial aspects of decision making **but** urban economics focuses **not only** on the location decisions of firms, **but also** of cities themselves **as** cities themselves represent centers of economic activity.

注意连词的含义：but 表示转折，not only...but also... 表示递进（不但……而且……），as 表示原因。

Exercises

[1] 将课文中下划线句子或段落译成中文。

[2] 读熟或背诵上述复杂长句。

[3] 解析下列单词[解析单词是指根据单词中包含的词根、前缀推导出单词的含义，例如，apathy（n. 漠然，冷淡），解析式为：a-（无）+ path（感情）+ y →无感情→冷漠，冷淡]。解析的单词选自上文"**New Words**"。

accessibility n. 可达性	immobile adj. 固定的，不变的，稳定的
commuter n. 乘公交车辆上下班者	macroeconomics n. 宏观经济学
conjunction n. 连接，联合	polycentricity n. 多中心，多中心性
deficit n. 赤字，亏损	preside v. 主持，指挥
fragmented adj. 成碎片的，片段的	synonymous adj. 同义的

Unit 19

Urban Geography

Urban geography is a **subdiscipline** of geography that **derives** from the study of cities and urban processes. Urban geographers and urbanists examine various aspects of urban life and the built environment. *Scholars, activists, and the public have participated in, studied, and* ***critiqued*** *flows of economic and natural resources, human and non–human bodies, patterns of development and infrastructure, political and institutional activities, governance, decay and renewal, and notions of socio–spatial* ***inclusions****,* ***exclusions****, and everyday life.*

1 Research interest

Urban geographers are primarily concerned with the ways in which cities and towns are constructed, governed and experienced. Alongside neighboring disciplines such as urban **anthropology**, urban planning and urban sociology, urban geography mostly investigates the impact of urban processes on the earth's surface's social and physical structures. Urban geographical research can be part of both human geography and physical geography.

The two fundamental aspects of cities and towns, from the geographic **perspective** are:

· Location ("systems of cities"): spatial distribution and the complex patterns of movement, flows and linkages that bind them in space; and

· Urban structure ("cities as systems"): study of patterns of distribution and interaction within cities, from **quantitative**, **qualitative**, structural, and behavioral perspectives.

2 Research topics

2.1 Cities as centers of manufacturing and services

Cities differ in their economic makeup, their social and **demographic** characteristics, and the roles they play within the city system. One can trace these differences back to regional variations in the local resources on which growth was based during the early development of the urban pattern and in part to the **subsequent** shifts in the competitive advantage of regions brought about by changing locational forces affecting regional specialization within the framework of a market economy. The recognition of different city types is **critical** for the classification of cities in urban geography. For such classification, emphasis is given in particular to functional town classification and the basic **underlying** dimensions of the city system.

The purpose of classifying cities is twofold. On the one hand, it is undertaken to search reality for hypotheses. In this context, the recognition of different types of cities on the basis of, for example, their functional specialization may enable the identification of spatial regularities in the distribution and structure of urban functions and the formulation of hypotheses about the resulting patterns. On the other hand, classification is undertaken to structure reality in order to test specific hypotheses that have already been formulated. For example, to test the hypotheses that cities with a diversified economy grow at a faster rate than those with a more specialized economic base, cities must first be classified so that diversified and specialized cities can be differentiated.

The simplest way to classify cities is to identify the distinctive role they play in the city system. There are three distinct roles:

· central places: functioning primarily as service centers for local hinterlands;

· transportation cities: performing **break-of-bulk** and allied functions for larger regions;

· specialized-function cities: dominated by one activity such as mining, manufacturing or recreation and serving national and international markets.

The composition of a city's labor force has traditionally been regarded as the best indicator of functional specialization, and different city types have been most frequently identified from the analysis of employment profiles. Specialization in a given activity is said to exist when employment in it exceeds some critical level.

The relationship between the city system and the development of manufacturing has become very apparent. The rapid growth and spread of cities within the **heartland-hinterland** framework after 1870 was conditioned to a large extent by industrial developments, and the **decentralization** of population within the urban system in recent years is related in large part to the movement of employment in manufacturing away from traditional industrial centers. Manufacturing is found in nearly all cities, but its importance is measured by the proportion of total earnings received by the inhabitants of an urban area. When 25 percent or more of the total earnings in an urban region derive from manufacturing, that urban area is **arbitrarily designated** as a manufacturing center.

The location of manufacturing is affected by **myriad** economic and non-economic factors, such as the nature of the material inputs, the factors of production, the market and transportation costs. Other important influences include **agglomeration** and **external** economies, public policy and personal preferences. Although it is difficult to evaluate precisely the effect of the market on the location of manufacturing activities, two considerations are involved:

- the nature of and demand for the product;
- transportation costs.

2.2 Urbanization

Urbanization, the **transformation** of population from rural to urban, is a major phenomenon of the modern era and a central topic of study.

3 History of the discipline

Urban geography arrived as a critical sub-discipline with the 1973 publication of David Harvey's *Social Justice and the City*, which was heavily influenced by previous work by Anne Buttimer. Prior to its emergence as its own discipline, urban geography served as the academic **extension** of what was otherwise a professional development and planning practice. *At the turn of the 19th century, urban planning began as a profession charged with* ***mitigating*** *the negative consequences of industrialization as documented by Friedrich Engels in his geographic analysis of the condition of the working class in England, 1844.*

In a 1924 study of urban geography, Marcel Aurousseau observed that urban geography cannot be considered a **subdivision** of geography because it plays such an important part. However, urban geography did emerge as a specialized discipline after World War II, amidst increasing urban planning and a shift away from the **primacy** of physical terrain in the study of geography. Chauncy Harris and Edward Ullman were among its earliest **exponents**.

Urban geography arose by the 1930s in the Soviet Union as an academic **complement** to active urbanization and communist urban planning, focusing on cities' economic roles and potential.

*Spatial analysis, behavioral analysis, Marxism, humanism, social theory, feminism, and post-modernism have arisen (in **approximately** this order) as overlapping **lenses** used within the field of urban geography in the West.*

Geographic information science, using digital processing of large data sets, has become widely used since the 1980s, with major applications for urban geography.

Notes

[1] 本文源自维基百科（From Wikipedia，the free encyclopedia，https://www.wikipedia.org/）。

[2] 本文主要相关文献：许学强，等，2009; Fyfe，et al，2005; Kaplan，et al，2003; Knox，et al，2005; LeGates，et al，2007; Pacione，2001。

[3] 课文中下划线部分为有关城市规划理论的重要论述，需强化阅读，每篇课文习题部分均要求将下划线句子或段落译成中文。

[4] 课文中下划线斜体字部分为复杂长句，其解析详见下文“**Complex Sentences**”。

New Words

agglomeration n. 成团，结块
anthropology n. 人类学〖词根 12. anthrop〗
approximately adv. 近似，大约
arbitrarily adv. 任意地，武断地
break-of-bulk n. 货物分卸
complement n. & v. 补充
critical adj. 批评的；关键的〖词根 48. crit〗
critique v. 批判，发表评论〖词根 48. crit〗
decay n. & v. 腐烂，腐朽
decentralization n. 分散〖前缀 18. de-; 词根 32. centr〗
demographic adj. 人口统计学的，人口统计的〖词根 54. dem; 词根 84. graph(y)〗
derive v. 得到，导出；源于，来自
designate v. 指明，指出〖前缀 18. de-; 词根 190. sign〗
exclusion n. 拒绝，排除〖前缀 26. ex-; 词根 40. clus〗
exponent n. 倡导者，拥护者

extension n. 伸展，扩大，延长〖前缀 26. ex-; 词根 208. tens〗

external adj. 外面的，外部的〖前缀 26. ex-〗

heartland n. 心脏地带，中心区域

hinterland n. 腹地，内地

inclusion n. 包含，内含物〖前缀 34. in-; 词根 40. clus〗

lens n. 透镜，镜头

mitigate v. 缓和，减轻

myriad adj. 无数的，各式各样的

perspective n. 观点，看法；洞察力；adj. 透视的〖前缀 53. per-; 词根 198. spec〗

primacy n. 首位，第一位；卓越〖173. prim〗

qualitative adj. 定性的

quantitative adj. 定量的

renewal n. 更新，重建〖前缀 61. re-〗

subdiscipline n.（学科的）分支〖前缀 65. sub-〗

subdivision n. 分支〖65. sub-〗

subsequent adj. 随后的，跟随的〖前缀 65. sub-; 词根 187. sequ〗

transformation n. 变化，转换〖前缀 71. trans-; 词根 76. form〗

underlying adj. 基础的，在下面的〖前缀 75. under-〗

Complex Sentences

[1] **Scholars, activists, and the public** *have participated in, studied, and critiqued* flows of economic and natural resources, human and non-human bodies, patterns of development and infrastructure, political and institutional activities, governance, decay and renewal, and notions of socio-spatial inclusions, exclusions, and everyday life.

黑体字部分为句子的主语，三个主语；斜体字部分为句子的谓语，三个谓语；下划线部分为句子的宾语，八个宾语。

[2] At the turn of the 19th century, **urban planning began as a profession** charged with mitigating the negative consequences of industrialization **as** documented by Friedrich Engels in his geographic analysis of the condition of the working class in England, 1844.

黑体字部分为句子的主干，charged 引导的过去分词短语作定语，修饰前面的 profession，as 引导的介词短语作定语，修饰前面的 the negative consequences of industrialization。

[3] **Spatial analysis, behavioral analysis, Marxism, humanism, social theory, feminism, and postmodernism have arisen** (in approximately this order) as overlapping lenses used within the field of urban geography in the West.

黑体字部分为句子的主语和谓语，主语很长，七个主语。as 引导的介词短语作状语，used 引导的过去分词短语作定语，修饰前面的 lenses，括号内容为解释性插入语。

Exercises

[1] 将课文中下划线句子或段落译成中文。

[2] 读熟或背诵上述复杂长句。

[3] 解析下列单词 [解析单词是指根据单词中包含的词根、前缀推导出单词的含义，例如，apathy（n. 漠然，冷淡），解析式为：a-（无）+ path（感情）+ y →无感情→冷漠，冷淡]。解析的单词选自上文“**New Words**”。

anthropology n. 人类学	inclusion n. 包含，内含物
decentralization n. 分散	perspective n. 观点，看法；洞察力；adj. 透视的
demographic adj. 人口统计（学）的	subsequent adj. 随后的，跟随的
designate v. 指明，指出	transformation n. 变化，转换
exclusion n. 拒绝，排除	underlying adj. 基础的，在下面的

Unit 20

Urban Ecology

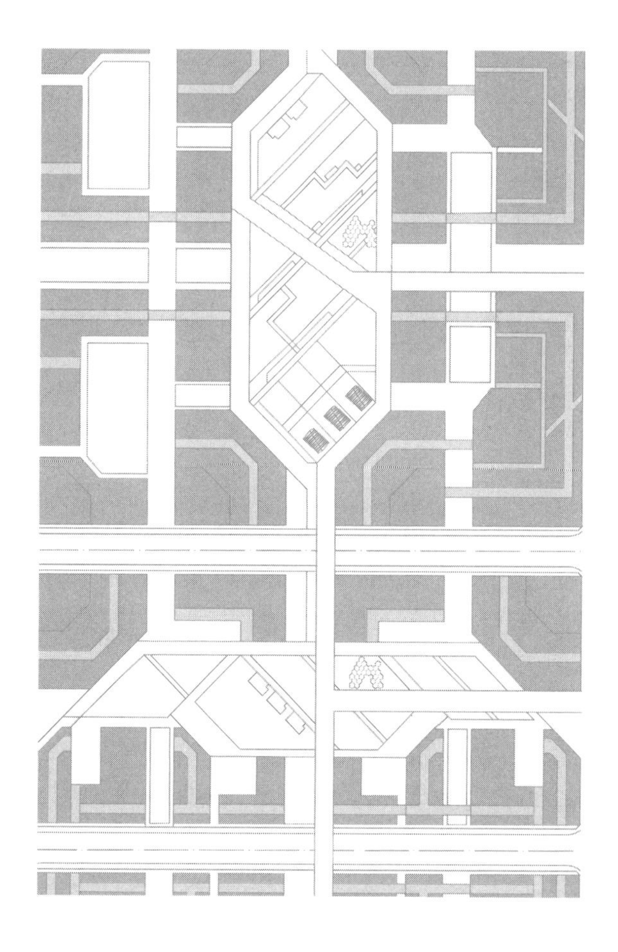

Urban **ecology** is the scientific study of the relation of living organisms with each other and their surroundings in the context of an urban environment. *The urban environment refers to environments dominated by high–density residential and commercial buildings, paved surfaces, and other urban–related factors that create a unique landscape* ***dissimilar*** *to most* ***previously*** *studied environments in the field of ecology.*

Urban ecology is a recent field of study compared to ecology as a whole. The study of urban ecology carries increasing importance because more than 50% of the world's population today lives in urban areas. At the same time, it is estimated that within the next forty years, two–thirds of the world's population will be living in expanding urban centers. The **ecological** processes in the urban environment are **comparable** to those outside the urban context. However, the types of urban habitats and the species that inhabit them are poorly documented. Often, explanations for phenomena examined in the urban setting because of urbanization are the center for scientific research.

1 History

Ecology has historically focused on **pristine** natural environments, but by the 1970s many **ecologists** began to turn their interest towards ecological **interactions** taking place in, and caused by urban environments. *Jean–Marie Pelt's 1977 book The Re–Naturalized Human, Brian Davis' 1978 publication Urbanization and the* ***Diversity*** *of Insects, and Sukopp et al.'s 1979 article The Soil,* ***Flora*** *and Vegetation of Berlin's Wasteland are some* *of the first publications to recognize the importance of urban ecology as a separate and distinct form of ecology the same way one might see landscape ecology as different from population ecology.* Forman and Godron's 1986 book *Landscape Ecology* first distinguished urban settings and landscapes from other landscapes by dividing all landscapes into five broad types. These types were divided by the intensity of human influence ranging from pristine natural environments to urban centers.

Urban ecology is recognized as a **diverse** and complex concept which differs in application between North America and Europe. The European concept of urban ecology examines the **biota** of urban areas, while the North American concept has traditionally examined the social sciences of the urban landscape, as well as the **ecosystem fluxes** and processes.

2 Methods

Since urban ecology is a **subfield** of ecology, many of the techniques are similar to that of ecology. Ecological study techniques have been developed over centuries, but many of the techniques used for urban ecology are more recently developed. Methods used for studying urban ecology **involve** chemical and **biochemical** techniques, temperature recording, heat mapping, remote sensing, and long-term ecological research sites.

2.1 Chemical and biochemical techniques

Chemical techniques may be used to determine **pollutant concentrations** and their effects. Tests can be as simple as dipping a manufactured test strip, as in the case of pH testing, or be more complex, as in the case of examining the spatial and temporal variation of heavy metal **contamination** due to industrial runoff. In that particular study, livers of birds from many regions of the North Sea were ground up and mercury was **extracted**. Additionally, mercury bound in feathers was extracted from both live birds and from museum specimens to test for mercury levels across many decades. Through these two different measurements, researchers were able to make a complex picture of the spread of mercury due to industrial runoff both spatially and temporally.

Other chemical techniques include tests for **nitrates**, **phosphates**, **sulfates**, etc. which are commonly associated with urban pollutants such as fertilizer and industrial **byproducts**. These biochemical fluxes are studied in the atmosphere (e.g. greenhouse gasses), **aquatic** ecosystems and soil vegetation. Broad reaching effects of these biochemical fluxes can be seen in various aspects of both the urban and surrounding rural ecosystems.

2.2 Temperature data and heat mapping

Temperature data can be used for various kinds of studies. An important aspect of temperature data is the ability to **correlate** temperature with various factors that may be affecting or occurring in the environment. Oftentimes, temperature data is collected long-term by the Office of Oceanic and Atmospheric Research (OAR), and made available to the scientific community through the National Oceanic and Atmospheric Administration (NOAA). Data can be overlaid with maps of terrain, urban features, and other spatial areas to create heat maps. These heat maps can be used to view trends and distribution over time and space.

2.3 Remote sensing

Remote sensing is the technique in which data is collected from distant locations through the use of satellite imaging, radar, and **aerial** photographs. In urban ecology, remote sensing is used to collect data about terrain, weather patterns, light, and vegetation. One application of remote sensing for urban ecology is to detect the productivity of an area by measuring the **photosynthetic** wavelengths of **emitted** light. Satellite images can also be used to detect differences in temperature and landscape diversity to detect the effects of urbanization.

2.4 LTERs and long-term data sets

Long-term ecological research (LTER) sites are research sites funded by the government that have collected reliable long-term data over an **extended** period of time in order to identify long-term climatic or ecological trends. These sites provide long-term temporal and spatial data such as average temperature, rainfall and other ecological processes. The main purpose of LTERs for urban ecologists is the collection of vast amounts of data over long periods of time. These long-term data sets can then be analyzed to find trends relating to the effects of the urban environment on various ecological processes, such as species diversity and abundance over time. Another example is the examination of temperature trends that are accompanied with the growth of urban centers.

3 Urban effects on the environment

Humans are the driving force behind urban ecology and influence the environment in a variety of ways, such as **modifying** land surfaces and waterways, introducing foreign species, and altering **biogeochemical** cycles. Some of these effects are more apparent, such as the **reversal** of the Chicago River to **accommodate** the growing pollution levels and trade on the river. Other effects can be more **gradual** such as the change in global climate due to urbanization.

3.1 Modification of land and waterways

Humans place high demand on land not only to build urban centers, but also to build surrounding suburban areas for housing. Land is also allocated for agriculture to **sustain** the growing population of the city. Expanding cities and suburban areas necessitate corresponding **deforestation** to meet the land–use and resource requirements of urbanization. Key examples of this are deforestation in the United States and Brazil.

Along with **manipulation** of land to suit human needs, natural water resources such as rivers and streams are also modified in urban establishments. **Modification** can come in the form of dams, artificial canals, and even the reversal of rivers. **Reversing** the flow of the Chicago River is a major example of urban environmental modification. Urban areas in natural desert settings often bring in water from far areas to maintain the human population and will likely have effects on the local desert climate. Modification of aquatic systems in urban areas also results in decreased stream diversity and increased pollution.

3.2 Trade, shipping, and spread of invasive species

Both local shipping and long–distance trade are required to meet the resource demands important in maintaining urban areas. **Carbon dioxide emissions** from the transport of goods also contribute to **accumulating** greenhouse gases and nutrient **deposits** in the soil and air of urban environments. In addition, shipping **facilitates** the **unintentional** spread of living organisms, and introduces them to environments that they would not naturally inhabit. Introduced or **alien** species are populations of organisms living in a range in which they did not naturally **evolve** due to **intentional** or **inadvertent** human activity. Increased transportation between urban centers furthers the incidental movement of animal and plant species. Alien species often have no natural **predators** and **pose** a **substantial** threat to the dynamics of existing ecological populations in the new environment where they are introduced. Such **invasive** species are **numerous** and include house sparrows, ring–necked pheasants, European starlings, brown rats, Asian carp, American bullfrogs, emerald ash borer, kudzu vines, and zebra mussels among numerous others, most notably **domesticated** animals. In Australia, it has been found that removing Lantana (L. camara, an alien species) from urban greenspaces can surprisingly have negative impacts on bird diversity locally, as it provides **refugia** for species like the superb fairy (Malurus cyaneus) and silvereye (Zosterops lateralis), in the absence of native plant **equivalents**. Although, there seems to be a density **threshold** in which too much Lantana (thus **homogeneity** in vegetation cover) can lead to a decrease in bird species richness or abundance.

3.3 Human effects on biogeochemical pathways

Urbanization results in a large demand for chemical use by industry, construction, agriculture, and energy providing services. Such demands have a substantial impact on biogeochemical cycles, resulting in phenomena such as **acid** rain, **eutrophication**, and global warming. Furthermore, natural biogeochemical cycles in the urban environment can be **impeded** due to **impermeable** surfaces that prevent nutrients from returning to the soil, water, and atmosphere.

Demand for fertilizers to meet agricultural needs **exerted** by expanding urban centers can alter chemical **composition** of soil. Such effects often result in **abnormally** high concentrations of **compounds** including **sulfur**, **phosphorus**, **nitrogen**, and heavy metals. In addition, nitrogen and phosphorus used in fertilizers have caused severe problems in the form of agricultural **runoff**, which alters the concentration of these compounds in local rivers and streams, often resulting in **adverse** effects on native species. A well–known effect of agricultural runoff is the phenomenon of eutrophication. When the fertilizer chemicals from agricultural runoff reach the ocean, an **algal bloom** results, then rapidly dies off. The dead **algae biomass** is **decomposed** by bacteria that also consume large quantities of oxygen, which they obtain from the water, creating a "dead zone" without oxygen for fish or other organisms. A classic example is the dead zone in the Gulf of Mexico due to agricultural runoff into the Mississippi River.

Just as pollutants and alterations in the biogeochemical cycle alter river and ocean ecosystems, they exert likewise effects in the air. Smog **stems** from the accumulation of chemicals and pollution and often **manifests** in urban settings, which has a great impact on local plants and animals. Because urban centers are often considered point sources for pollution, unsurprisingly local plants have adapted to **withstand** such conditions.

4 Urban effects on climate

Urban environments and **outlying** areas have been found to exhibit **unique** local temperatures, **precipitation**, and other characteristic activity due to a variety of factors such as pollution and altered geochemical cycles. Some examples of the urban effects on climate are urban heat island, **oasis** effect, greenhouse gases, and acid rain. This further stirs the debate as to whether urban areas should be considered a unique **biome**. Despite common trends among all urban centers, the surrounding local environment heavily influences much of the climate. One such example of regional differences can be seen through the urban heat island and oasis effect.

4.1 Urban heat island effect

The urban heat island is a phenomenon in which central regions of urban centers exhibit higher mean temperatures than surrounding urban areas. Much of this effect can be **attributed** to low city **albedo**, the reflecting power of a surface, and the increased surface area of buildings to absorb solar radiation. **Concrete**, **cement**, and metal surfaces in urban areas tend to absorb heat energy rather than reflect it, contributing to higher urban temperatures. Brazel et al. found that the urban heat island effect demonstrates a positive correlation with population density in the city of Baltimore. The heat island effect has corresponding ecological consequences on resident species. However, this effect has only been seen in **temperate** climates.

4.2 Greenhouse gases

Greenhouse gas emissions include those of carbon dioxide and **methane** from the **combustion** of fossil fuels to supply energy needed by vast urban **metropolises**. Other greenhouse gases include water vapor, and **nitrous oxide**. Increases in greenhouse gases due to urban transport, construction, industry and other demands have been correlated strongly with increase in temperature. Sources of methane are agricultural **dairy** cows and **landfills**.

4.3 Acid rain and pollution

Processes related to urban areas result in the emission of numerous pollutants, which change corresponding nutrient cycles of carbon, sulfur, nitrogen, and other elements. Ecosystems in and around the urban center are especially influenced by these point sources of pollution. High sulfur dioxide concentrations resulting from the industrial demands of urbanization cause rainwater to become more **acidic**. Such an effect has been found to have a significant influence on locally affected populations, especially in aquatic environments. Wastes from urban centers, especially large urban centers in developed nations, can drive biogeochemical cycles on a global scale.

4.4 Urban environment as an anthropogenic biome

The urban environment has been classified as an **anthropogenic** biome, which is characterized by the **predominance** of certain species and climate trends such as urban heat island across many urban areas. Examples of species characteristic of many urban environments include, cats, dogs, mosquitoes, rats, flies, and pigeons, which are all generalists. Many of these are dependent on human activity and have adapted accordingly to the **niche** created by urban centers.

5 Biodiversity and urbanization

Research thus far indicates that, on a small scale, urbanization often increases the **biodiversity** of **non-native** species while reducing that of **native** species. This **normally** results in an overall reduction in species **richness** and increase in total biomass and species **abundance**. Urbanization also reduces diversity on a large scale.

Urban stream **syndrome** is a consistently observed trait of urbanization characterized by high nutrient and **contaminant** concentration, altered stream **morphology**, increased dominance of dominant species, and decreased biodiversity. The two primary causes of urban stream syndrome are storm water runoff and wastewater treatment plant **effluent**.

5.1 Changes in diversity

Diversity is normally reduced at **intermediate-low** levels of urbanization but is always reduced at high levels of urbanization. These effects have been observed in **vertebrates** and **invertebrates** while plant species tend to increase with intermediate-low levels of urbanization but these general trends do not apply to all organisms within those groups. For example, McKinney's (2006) review did not include the effects of urbanization on fishes and of the 58 studies on invertebrates, 52 included insects while only 10 included spiders. There is also a geographical bias as most of the studies either took place in North America or Europe.

The effects of urbanization also depend on the type and range of resources used by the organism. Generalist species, those that use a wide range of resources and can thrive under a large range of living conditions, are likely **survive** in uniform environments. Specialist species, those that use a narrow range of resources and can only cope with a narrow range of living conditions, are unlikely to cope with uniform environments. There will likely be a variable effect on these two groups of organisms as urbanization alters habitat uniformity.

5.2 Cause of diversity change

The urban environment can decrease diversity through **habitat** removal and species

homogenization—the increasing similarity between two previously distinct biological communities. Habitat **degradation** and habitat **fragmentation** reduces the amount of suitable habitat by urban development and separates suitable patches by **inhospitable** terrain such as roads, neighborhoods, and open parks. Although this replacement of suitable habitat with unsuitable habitat will result in **extinctions** of native species, some shelter may be **artificially** created and promote the survival of non–native species (e.g. house sparrow and house mice nests). Urbanization promotes species homogenization through the extinction of **native endemic species** and the introduction of non–native species that already have a widespread abundance. Changes to the habitat may promote both the extinction of native endemic species and the introduction of non–native species. The effects of habitat change will likely be similar in all urban environments as urban environments are all built to **cater** to the needs of humans.

The urban environment can also increase diversity in a number of ways. Many foreign organisms are introduced and **dispersed** naturally or artificially in urban areas. Artificial introductions may be intentional, where organisms have some form of human use, or **accidental**, where organisms **attach** themselves to transportation vehicles. Humans provide food sources (e.g. birdfeeder seeds, trash, garden compost) and reduce the numbers of large natural **predators** in urban environments, allowing large populations to be supported where food and predation would normally limit the population size. There are a variety of different habitats available within the urban environment as a result of differences in land use allowing for more species to be supported than by more uniform habitats.

6 Civil engineering and sustainability

Cities should be planned and constructed in such a way that **minimizes** the urban effects on the surrounding environment (urban heat island, precipitation, etc.) as well as **optimizing** ecological activity. For example, increasing the albedo, or **reflective** power, of surfaces in urban areas, can minimize urban heat island, resulting in a lower **magnitude** of the urban heat island effect in urban areas. By minimizing these **abnormal** temperature trends and others, ecological activity would likely be improved in the urban setting.

6.1 Need for remediation

Urbanization has indeed had a profound effect on the environment, on both local and global scales. Difficulties in actively constructing habitat **corridor** and returning biogeochemical cycles to **normal** raise the question as to whether such goals are **feasible**. However, some groups are working to return areas of land affected by the urban landscape to a more natural state. This includes using landscape architecture to model natural systems and restore rivers to pre-urban states.

6.2 Sustainability

With the ever-increasing demands for resources necessitated by urbanization, recent campaigns to move toward sustainable energy and resource consumption, such as LEED certification of buildings, Energy Star certified appliances, and zero emission vehicles, have gained **momentum**. Sustainability reflects techniques and consumption ensuring reasonably low resource use as a component of urban ecology. Techniques such as carbon **recapture** may also be used to **sequester** carbon compounds produced in urban centers rather continually emitting more of the greenhouse gas.

7 Summary

Urbanization results in a series of both local and far-reaching effects on biodiversity, biogeochemical cycles, **hydrology**, and climate, among many other stresses. Many of these effects are not fully understood, as urban ecology has only recently **emerged** as a scientific discipline and much more research remains to be done. Research on cities outside the US and Europe remains limited. Observations on the impact of urbanization on biodiversity and species interactions are consistent across many studies but **definitive** mechanisms have yet to be established. Urban ecology **constitutes** an important and highly relevant subfield of ecology, and further study must be **pursued** to more fully understand the effects of human urban areas on the environment.

Notes

[1] 本文源自维基百科（From Wikipedia，the free encyclopedia，https://www.wikipedia.org/）。

[2] 本文主要相关文献：沈清基，2011，2019；宋永昌，等，2000；Niemela，1999；Roseland，1997。

[3] 课文中下划线部分为有关城市规划理论的重要论述，需强化阅读，每篇课文习题部分均要求将下划线句子或段落译成中文。

[4] 课文中下划线斜体字部分为复杂长句，其解析详见下文“**Complex Sentences**”。

New Words

abnormal adj. 反常的，异常的〖前缀 2. ab-；词根 138. norm〗

abnormally adv. 不正常地 〖前缀 2. ab-; 词根 138. norm〗

abundance n. 多度

accidental adj. 意外的，偶然的

accommodate v. 容纳，向……提供；使适应，使和谐一致

accumulate n. 堆积，积累 〖前缀 3. ac-; 词根 51. cumul〗

acid adj. 酸的，酸性的，尖刻的 〖词根 1. acid〗

acidic adj. 酸的，酸性的 〖词根 1. acid〗

adverse adj. 不利的，有害的 〖前缀 3. ad-; 词根 226. vers〗

aerial adj. 空气的，空中的，航空的 〖词根 3. aer〗

albedo n. 反照率

algae n. 藻类

algal adj. 海藻的

alien adj. 外国的，异己的；n. 外国人，外星人 〖词根 6. alien〗

anthropogenic adj. 人类起源论的，人为的 〖词根 12. anthrop; 词根 81. gen〗

aquatic adj. 水生的，水产的 〖词根 13. aqu〗

artificially adv. 人工地，不自然地 〖词根 15. art; 词根 65. fic〗

attach v. 系，贴，附加 〖词根 204. tach〗

attribute v. 认为……是，把……归于 〖词根 214. tribut〗

biochemical adj. 生物化学的 〖词根 22. bio〗

biodiversity n. 生物多样性 〖词根 22. bio; 前缀 19. di-; 词根 226. vers〗

biogeochemical adj. 生物地球化学的 〖词根 22. bio; 词根 geo, 地〗

biomass n. 生物量 〖词根 22. bio〗

biome n. 生物群落 〖词根 22. bio〗

biota n. 生物区系，生物群 〖词根 22. bio〗

bloom v. 开花，植物繁盛

byproduct n. 副产品，副作用 〖前缀 13. by-〗

carbon n. 碳

cater v. 迎合，满足需要

cement n. 水泥

combustion n. 燃烧

comparable adj. 可比较的，比得上的

composition n. 组成，构成 〖前缀 16. com-; 词根 170. pos〗

compound n. 复合物

comprise v. 包含，包括，由……组成

concentration n. 集中；浓度 〖前缀 16. con-; 词根 32. centr〗

concrete n. 混凝土

constitute v. 构成，组成；制定，设立

contaminant n. 污染物

contamination n. 污染，弄脏

correlate v. 联系，使互相关联

corridor n. 走廊，通道

dairy n. 牛奶场，乳品店

decompose v. 分解，腐烂〖前缀 18. de–; 前缀 16. com–; 词根 170. pos〗

definitive adj. 最后的，确定的〖前缀 18. de–; 词根 68. fin〗

deforestation n. 采伐森林〖前缀 18. de–〗

degradation n. 堕落，恶化〖前缀 18. de–; 词根 grad, 等级〗

deposit n. 储蓄，存款；沉淀物〖前缀 18. de–; 词根 170. pos〗

dioxide n. 二氧化物〖前缀 19. di–〗

disperse v. 分散，散开，散播〖前缀 20. dis–〗

dissimilar adj. 不同的，不相似的〖前缀 20. dis–; 词根 191. simil〗

diverse adj. 不同的，多种多样的〖前缀 19. di–; 词根 226. vers〗

diversity n. 多样性〖前缀 19. di–; 词根 226. vers〗

domesticate v. 驯养

ecological adj. 生态的，生态学的〖词根 57. eco〗

ecologist n. 生态学者，生态学家〖词根 57. eco〗

ecology n. 生态学〖词根 57. eco〗

ecosystem n. 生态系统〖词根 57. eco〗

effluent n. 污水，流出的水流；adj. 流出的，发出的〖前缀 ef–(= 26. ex–); 词根 75. flu〗

emerge v. 出现，浮现

emission n. 排放，散发〖前缀 e–(= 26. ex–); 词根 129. miss〗

emit v. 发出，发射〖前缀 e–(= 26. ex–); 词根 129. mit〗

endemic adj. 某地特有的，地方性的

equivalent adj. 相等的，相当的；n. 对等物〖词根 58. equ〗

eutrophication n. 富营养化〖前缀 25. eu–〗

evolve v. 发展，进化〖词根 234. volv〗

exert v. 发挥，运用

extend v. 延伸，扩大，推广〖前缀 26. ex–; 词根 208. tend〗

extinction n. 消灭，灭绝

extract v. 提取，拔出，勒索〖前缀 26. ex–; 词根 213. tract〗

facilitate v. 促进，助长

feasible adj. 可行的

flora n. 植物群，植物区系〖词根 74. flor〗

flux n. 流量，流出〖词根 75. flu〗

fragmentation n. 分裂，破碎〖词根 77. frag〗

generalist n. 通才，多面手

habitat n. 栖息地，住处

homogeneity n. 一致性，均匀性〖前缀 30. homo–; 词根 81. gen〗

homogenization n. 均匀化，均质化〖前缀 30. homo–; 词根 81. gen〗

hydrology n. 水文学，水文地理学〖词根 95. hydr〗

impede v. 阻碍，妨碍，阻止〖前缀 33. im–; 词根 156. ped〗

impermeable adj. 不可渗透的〖前缀 33. im–; 前缀 53. per–; 〗

inadvertent adj. 漫不经心的〖前缀 34. in–; 前缀 3. ad–; 词根 226. vert〗

inhospitable adj. 不好客的，不友好的，不适于居住的〖前缀 34. in–〗

intentional adj. 有意的，故意的，策划的

interaction n. 相互作用，相互影响〖前缀 35. inter–; 2. act〗

intermediate adj. 中间的，中级的〖前缀 35. inter–; 词根 122. medi〗

invasive adj. 侵略性的，扩散的〖前缀 34. in–; 词根 219. vas〗

invertebrate n. 无脊椎动物；adj. 无脊椎的〖前缀 34. in–〗

involve v. 包含，牵扯，使参与〖前缀 34. in–; 词根 234. volv〗

landfill n. 垃圾填筑池，废渣填埋法

magnitude n. 巨大，重大，重要〖词根 118. magn〗

manifest v. 显示，表明，证明；adj. 明白的，明显的

manipulation n. 操作，操纵，控制〖词根 119. man〗

methane n. 甲烷，沼气

metropolises n. 一国的主要城市（不一定是首都）

minimize v. 把……减至最低程度

modification n. 修改，修正

modify v. 修改，修正

momentum n. 动量，势头

morphology n. 形态学〖词根 131. morph〗

native adj. 土著的，天生的，本国的

nitrate n. 硝酸盐

nitrogen n. 氮

nitrous adj. 氮的，含氮的

non-native adj. 非原生的，非本地的〖前缀 45. non-〗

normal adj. 正常的，正规的〖词根 138. norm〗

normally adv. 正常地，通常地〖词根 138. norm〗

numerous adj. 很多的，许多的〖词根 142. numer〗

oasis n. 绿洲

optimize v. 使最优化

outlying adj. 偏僻的，边远的〖前缀 47. out-〗

oxide n. 氧化物

phosphate n. 磷酸盐

phosphorus n. 磷

photosynthetic adj. 光合作用的，促进光合作用的〖词根 photo, 光；前缀 70. syn-〗

pollutant n. 污染物

pose v. 提出，造成，引起，产生〖词根 170. pos〗

precipitation n. 降水量，降雨量〖词根 36. cip〗

predator n. 食肉动物，捕食其他动物的动物

predominance n. 优势，主导或支配的地位

previously adv. 事先，以前〖前缀 57. pre-〗

pristine adj. 太古的，原始状态的；纯朴的，纯洁的〖词根 173. pri〗

pursue v. 继续，追求

recapture v. 重新捕获〖前缀 61. re-; 词根 26. cap〗

reflective adj. 反射的

refuge n. 避难，庇护，躲避

reversal n. 反向，反转，倒转〖前缀 61. re-; 词根 226. vers〗

reverse v. 反转，颠倒〖前缀 61. re-; 词根 226. vers〗

richness n. 丰富度

runoff n. 径流

sequester v. 使隔绝，使隔离

specialist n. 专家，行家

stem n. 树干，叶梗，叶茎；v. 起源，发生

subfield n. 分支〖前缀 65. sub-〗

subset n. 子集〖前缀 65. sub-〗

substantial adj. 大量的

sulfate n. 硫酸盐

sulfur n. 硫

survive v. 幸存〖前缀 69. sur–; 词根 231. viv〗
sustain v. 维持，支撑，支持
syndrome n. 综合症状
temperate adj. 有节制的，适度的，(气候)温和的〖词根 206. tem〗
threshold n. 阈值，临界值
unintentional adj. 无意的，无心的〖前缀 74. un–〗
vertebrate n. 脊椎动物；adj. 有脊椎的
withstand v. 经受，承受〖前缀 78. with–〗

Complex Sentences

[1] **The urban environment refers to environments** dominated by high–density residential and commercial buildings, paved surfaces, and other urban–related factors **that** create a unique landscape dissimilar to most previously studied environments in the field of ecology.
黑体字部分为句子的主干（主语、谓语与宾语）；dominated 引导的过去分词短语作定语，修饰前面的 environments；that 引导的从句为定语从句，亦修饰前面的 environment，其中的 dissimilar to most previously studied environments in the field of ecology 形容词短语，作定语，修饰前面的 a unique landscape。

[2] *Jean–Marie Pelt's 1977 book The Re–Naturalized Human, Brian Davis' 1978 publication Urbanization and the Diversity of Insects, and Sukopp et al.'s 1979 article The Soil, Flora and Vegetation of Berlin's Wasteland are some of the first publications* to recognize the importance of urban ecology as a separate and distinct form of ecology **the same way** one might see landscape ecology as different from population ecology.
斜体字部分为句子的主干（主语、谓语与表语），本句的主语很长。动词不定式短语 to recognize the importance of urban ecology as a separate and distinct form of ecology 作定语，修饰前面的 some of the first publications; the same way 引导的从句为状语从句。

Exercises

[1] 将课文中下划线句子或段落译成中文。
[2] 读熟或背诵上述复杂长句。
[3] 解析下列单词[解析单词是指根据单词中包含的词根、前缀推导出单词的含义，例如，apathy（n. 漠然，冷淡），解析式为：a–（无）+ path（感情）+ y →无感情→冷漠，冷淡]。解析的单词选自上文“**New Words**”。

abnormal adj. 反常的，异常的	impede v. 阻碍，妨碍，阻止
aerial adj. 空气的，空中的，航空的	impermeable adj. 不可渗透的
anthropogenic adj. 人类起源论的，人为的	invasive adj. 侵略性的，扩散的
biodiversity n. 生物多样性	magnitude n. 巨大，重大，重要
composition n. 组成，构成	manipulation n. 操作，操纵，控制
decompose v. 分解，腐烂	morphology n. 形态学
deposit n. 储蓄，存款；沉淀物	precipitation n. 降水量，降雨量
equivalent adj. 相等的，相当的；n. 对等物	previously adv. 事先，以前
extract v. 提取，拔出，勒索	reversal n. 反向，反转，倒转
homogeneity n. 一致性，均匀性	reverse v. 反转，颠倒

Bibliography and References

上篇

[1] 蔡瑞，孟醒 . ACT 必备核心词汇 [M]. 杭州：浙江教育出版社，2015.

[2] 蒋争 . 英语词汇的奥秘 [M]. 北京：中国国际广播出版社，2000.

[3] 卡罗尔・沃克 . 朗文常用英文词根词典 [M]. 北京：外语教学与研究出版社，1999.

[4] 刘毅 . 英文字根字典 [M]. 北京：外文出版社，2010.

[5] 阎传海 . GRE 词汇分类记忆宝典 [M]. 北京 : 世界图书出版公司北京公司，2013.

[6] 俞敏洪 . GRE 词汇精选 [M]. 北京：群言出版社，2008.

[7] 俞敏洪 . NEW GRE 词汇精选 [M]. 北京：群言出版社，2011.

[8] 俞敏洪 . GMAT 词汇精选 [M]. 西安：西安交通大学出版社，2014.

[9] 俞敏洪 . SAT 词汇 [M]. 北京：群言出版社，2014.

[10] 俞敏洪 . TOEFL 词汇 [M]. 西安：西安交通大学出版社，2012.

[11] 俞敏洪 . TOEFL 词汇（45 天突破版）[M]. 北京：群言出版社，2011.

[12] 俞敏洪 . 雅思词汇（加强版）[M]. 西安：西安交通大学出版社，2014.

[13] 张一冰 . SAT 经典词汇 [M]. 上海：上海译文出版社，2016.

下篇

[14] 常乐，孙元元 . 建筑英语 [M]. 北京：外语教学与研究出版社，2008.

[15] 靳慧霞 . 建筑学专业英语 [M]. 北京：化学工业出版社，2016.

[16] 李明章 . 建筑类专业英语（第三册）[M]. 北京：中国建筑工业出版社，1997.

[17] 武涛，杨滨章 . 风景园林专业英语 [M]. 重庆：重庆大学出版社，2012.

[18] 徐良，汪丽君，舒平 . 建筑专业英语学习教程 [M]. 天津：天津大学出版社，2008.

[19] 徐铁城，王庆昌 . 建筑类专业英语（第二册）[M]. 北京：中国建筑工业出版社，1997.

[20] 阎传海 . GRE 句子填空对策与技巧 [M]. 北京 : 世界图书出版公司北京公司，2016.

[21] 杨鹏 . GRE & GMAT 阅读难句教程 [M]. 杭州：浙江教育出版社，2015.

[22] 张冠增 . 城市规划专业英语教程 [M]. 北京：中国建筑工业出版社，2010.

[23] 赵纪军，陈晓彤 . 城乡规划专业英语 [M]. 武汉：华中科技大学出版社，2016.

[24] 郑启颖 . 建筑专业英语 [M]. 武汉：华中科技大学出版社，2014.

[25] 彼得·卡尔索普，温锋华，彭卓见 . 新城市主义在中国的实践与未来 [J]. 北京规划建设，2019，（5）：191–196.

[26] 方澜，于涛方，钱欣 . 战后西方城市规划理论的流变 [J]. 城市问题，2002，（1）：10–13.

[27] 顾朝林 . 战后西方城市研究的学派 [J]. 地理学报，1994，49（4）：371–382.

[28] 顾朝林，刘佳燕等 . 城市社会学 [M]. 2 版 . 北京：清华大学出版社，2013.

[29] 金经元 . 再谈霍华德的明日的田园城市 [J]. 国外城市规划，1996，（4）：31–36.

[30] 李德华 . 城市规划原理 [M]. 3 版 . 北京：中国建筑工业出版社，2001.

[31] 栾峰 . 战后西方城市规划理论的发展演变与核心内涵——读 Nigel Taylor 的《1945 年以来的城市规划理论》[J]. 城市规划汇刊，2004，（6）：83–87.

[32] 栾峰 . 城市经济学 [M]. 北京：中国建筑工业出版社，2012.

[33] 马交国，孙海军 . 新城市主义引导下区域性综合交通枢纽规划实践审视——以济南新东站 TOD 规划设计为例 [J]. 科学与管理，2018，（4）：61–65.

[34] 彭秀良 . 雄安新区：理想愿景或为中国式田园城市 [J]. 北京规划建设，2017，（3）：180–182.

[35] 仇保兴 . 19 世纪以来西方城市规划理论演变的六次转折 [J]. 规划师，2003，19（11）：5–10.

[36] 沈清基 . 城市生态环境：原理、方法与优化 [M]. 北京：中国建筑工业出版社，2011.

[37] 沈清基，彭姗妮，慈海 . 现代中国城市生态规划演进与展望 [J]. 国际城市规划，2019，（4）：37–48.

[38] 沈玉麟 . 外国城市建设史 [M]. 北京：中国建筑出版社，1989.

[39] 宋永昌，由文辉，王祥荣 . 城市生态学 [M]. 上海：华东师范大学出版社，2000.

[40] 孙施文 . 现代城市规划理论 [M]. 北京：中国建筑工业出版社，2007.

[41] 唐子来 . 田园城市理念对于西方战后城市规划的影响 [J]. 城市规划会刊，1998，（6）：5–7，39.

[42] 王凯 . 从西方规划理论看我国规划理论建设之不足 [J]. 城市规划，2003，27（6）：66–71.

[43] 伍美琴，吴缚龙 . 西方规划理论对中国城市规划的启示 [J]. 国外城市规划，1994，（4）：14–19.

[44] 吴志强 . 百年现代城市规划中不变的精神和责任 [J]. 城市规划，1999，（1）：27–32.

[45] 吴志强 .《百年西方城市规划理论史纲》导论 [J]. 城市规划汇刊，2000，（2）：9–18，53.

[46] 吴志强，李德华 . 城市规划原理 [M]. 4 版 . 北京：中国建筑工业出版社，2010.

[47] 许学强，周一星，宁越敏 . 城市地理学 [M]. 2 版 . 北京：高等教育出版社，2009.

[48] 于泓 . Davidoff 的倡导性城市规划理论 [J]. 国外城市规划，2000，(1): 30–33.

[49] 张京祥 . 西方城市规划思想史纲 [M]. 南京：东南大学出版社，2005.

[50] 张庭伟 . 从“向权力讲授真理”到“参与决策权力”——当前美国规划理论界的一个动向 :“联络性规划”[J]. 城市规划，1999，23 (6): 33–36.

[51] 张衔春，胡国华 . 美国新城市主义运动：发展、批判与反思 [J]. 国际城市规划，2016，(3): 40–48.

[52] 张雅鹏 . 新城市主义理论对京津冀都市圈下小城镇的发展借鉴 [J]. 小城镇建设，2017，(4): 38–43.

[53] 周国艳，于立 . 西方现代城市规划理论概论 [M]. 南京：东南大学出版社，2010.

[54] 周岚 . 西方城市规划理论发展对中国之启迪 [J]. 国外城市规划，2001，(1): 34–37.

[55] 宗仁 . 霍华德“田园城市”理论对中国城市发展的现实借鉴 [J]. 现代城市研究，2018，(2): 77–81.

[56] Alexander, C. 1965. A city is not a tree. *Architectural Forum*, Vol. 122, no. 1. 严小婴 , 译 . 城市并非树形 [J]. 建筑师，1985，(24) .

[57] Arnstein, S. R. A ladder of citizen participation[J]. *Journal of the American Institute of Planners*, 1969, Vol. 35, July.

[58] Bacon, E. *Design of Cities*[M]. New York: Viking Press, 1967.

[59] Beevers, R. *The Garden City Utopia: A Critical Biography of Ebenezer Howard*[M]. Basingstoke: Macmillan, 1987.

[60] Bruton, M. J. (editor). *The Spirit and Purpose of Planning*[M]. London: Hutchinson, 1974.

[61] Buchanan, C. D. et al. *Traffic in Towns*[M]. London: HMSO, 1963 (shorted edition published by Penguin, Harmondsworth, 1964).

[62] Calthorpe, P. and W. Fulton. *The Regional City: Planning for the End of Sprawl*[M]. Washington, DC: Island Press, 2001.

[63] Castells, M. *The Urban Question: A Marxist Approach*[M]. London: Edward Arnold, 1977.

[64] Chadwick, G. F. *A Systems View of Planning*[M]. Oxford: Pergamon Press, 1971.

[65] Cherry, G. *Pioneers in British Town Planning*[M]. London: Architectural Press, 1981.

[66] Creese, W. *The Search for Environment: The Garden City Before and After*[M]. 2nd edition. New Haven, CT: Yale University Press, 1992.

[67] Davidoff, P. Advocacy and pluralism in planning[J]. *Journal of the American Institute of Planners*, 1965, Vol. 31, November.

[68] Duany, A. and E. PlaterZyberk. *Suburban Nation: The Rise of Sprawl and the Decline of the American Dream*[M]. New York: North Point Press, 2000.

[69] Dutton, J. A. *New American Urbanism: Re–forming the Suburban Metropolis*[M]. Milan: Skira, 2000.

[70] Etzioni, A. Mixed–scanning: a “third” approach to decision–making[J]. *Public Administration Review*,

1967.

[71] Faludi, A. *Planning Theory*[M]. Oxford: Pergamon Press, 1973.

[72] Faludi, A. *A Reader in Planning Theory*[M]. Oxford: Pergamon Press, 1973.

[73] Fishman, R. *Urban Utopias in the Twentieth Century: Ebenezer Howard, Frank Lloyd Wright and Le Corbusier*[M]. New York: Basic Books, 1977.

[74] Forester, J. Planning in the face of conflict[J]. *Journal of the American Planning Association*, 1987, 53, 3 (Summer).

[75] Forester, J. *Planning in the Face of Power*[M]. University of California Press, 1989.

[76] Forester, J. *Critical Theory, Public Policy, and Planning Practice: Toward a Critical Pragmatism*[M]. Albany: State University of New York Press, 1993.

[77] Forester, J. *The Deliberative Practitioner: Encouraging Participatory Planning Processes*[M]. Cambridge, MA: MIT Press, 1999.

[78] Friedmann, J. and C. Weaver. *Territory and Function*[M]. London: Edward Arnold, 1979.

[79] Friedmann, J. *Planning in the Public Domain: From Knowledge to Action*[M]. Princeton, NJ: Princeton University Press, 1988.

[80] Fyfe, N. and J. Kenny. *The Urban Geography Reader*[M]. London and New York: Routledge, 2005.

[81] Galloway, T. G. and Mahayni, R. G. Planning theory in retrospect: the process of paradigm change[J]. *Journal of the American Institute of Planners*, 1977, Vol. 43, no. 1.

[82] Geddes, P. *Cities in Evo1ution: An introduction to the Town Planning Movement and the Study of Civics*[M]. London: Ernest Benn (1968 edition), 1915.

[83] Giddens, A. *The Consequences of Modernity*[M]. Oxford: Polity Press, 1990.

[84] Hall, P. *Urban and Regional Planning*[M]. 4th edition. London and New York: Routledge. 彼得·霍尔，2002. 城市和区域规划（原著第四版）[M]. 邹德慈，李浩，陈熳莎，译 . 北京：中国建筑工业出版社，2008.

[85] Hall, P. *Cities of Tomorrow: An Intellectual History of Urban Planning and Design in the Twentieth Century*[M]. 3rd edition. Oxford: Blackwell, 2002.

[86] Harvey, D. *Social Justice and the City*[M]. London: Edward Arnold, 1973.

[87] Healey, P. *Land Use Planning and the Mediation of Urban Change: The British Planning System in Practice*[M]. Cambridge: Cambridge University Press, 1988.

[88] Healy, P. *Collaborative Planning: Shaping Places in Fragmented Societies*[M]. Basingstoke: Macmillan, 1977.

[89] Howard, E. *Garden Cities of Tomorrow* (originally published as *Tomorrow: A Peaceful Path to Real Reform*)[M]. London: Faber & Faber, 1965. 霍华德 . 明日的田园城市 [M]. 金经元，译 . 北京：商务印书馆，2000.

[90] Innes, J. E. Planning theory's emerging paradigm: communicative action and interactive practice[J].

Journal of Planning Education and Research, 1995, Vol. 14, no. 3.

[91] Jacobs, A. and D. Appleyard. Toward an urban design manifesto[J]. *Journal of the American Planning Association,* 1987, 53, 1 (Winter).

[92] Jacobs, A. *Great Streets*[M]. Cambridge: M.I.T. Press, 1995.

[93] Jacobs, J. *The Death and Life of Great American Cities*[M]. New York: Random House. 简·雅各布斯, 1961. 美国大城市的死与生 [M]. 金衡山，译 . 南京：译林出版社，2005.

[94] Jackson, F. *Sir Raymond Unwin: Architect, Planner and Visionary*[M]. London: Zwemmer, 1985.

[95] Jencks, C. *The Language of Post–Modern Architecture*[M]. London: Academy Editions, 1991.

[96] Kaplan, D. H., J. O. Wheeler, and S. Holloway. *Urban Geography*[M]. New York, Wiley, 2003.

[97] Katz, P. *The New Urbanism: Toward an Architecture of Community*[M]. New York: McGraw–Hill, 1994.

[98] Keeble, L. *Principles and Practice of Town and Country Planning*[M]. London: The Estates Gazette, 1952.

[99] Knox, P. and L. McCarthy. *Urbanization: An Introduction to Urban Geography*[M]. 2nd edition. New York: Prentice Hall, 2005.

[100] Kuhn, T. *The structure of Scientific Revolutions*[M]. Chicago, Ill.: University of Chicago Press, 1962(2nd edition 1970).

[101] Le Corbusier. *Urbanisme*[M]. Paris: Editions Cres, 1924(English translation by Frederick Etchells, published as *The City of Tomorrow.* London: The Architectural Press. 1971) .

[102] Le Corbusier. *La Ville Radieuse*[M]. Paris: Vincent, Freal et Cie, 1933(English translation by Pamela Knight, Eleanor Levieux and Derex Coltman, published as *The Radiant City*. London: Faber & Faber. 1967).

[103] LeGates, R., 张庭伟 . 为中国规划师的西方城市规划文献导读 [J]. 城市规划学刊，2007，（4）: 17–35.

[104] Lichfield, N., Kettle, P. and Whitbread, M. *Evaluation in the Planning Process*[M]. Oxford: Pergamon Press, 1975.

[105] Lin, J. and C. Mele. *The Urban Sociology Reader*[M]. London and New York: Routledge, 2004.

[106] Lindblom, C. E. The science of "muddling through" [J]. *Public Administration Review*, Spring, 1959.

[107] Lynch, K. A. *The Image of the City*[M]. Cambridge: M.I.T. Press, 1961.

[108] Lynch, K. A. *Site Planning*[M]. 3rd edition. Cambridge: M.I.T. Press, 1984.

[109] Lynch, K. A. *Good City Form*[M]. Cambridge: M.I.T. Press, 1988.

[110] Macdonald, E and M. Larice. *The Urban Design Reader*[M]. London and New York: Routledge, 2007.

[111] Macionis, J. and V. Parillo. *Cities and Urban Life*[M]. 4th Edition. New York: Prentice–Hall, 2006.

[112] McHarg, I. *Design with Nature*[M]. New York: Natural History Press, 1969.

[113] McLoughlin, J. B. *Urban and Regional Planning: A Systems Approach*[M]. London: Faber & Faber, 1969.

[114] Meller, H. *Patrick Geddes: Social Evolutionist and City Planner*[M]. London: Routledge, 1990.

[115] Mumford, L. *The Culture of Cities*[M]. New York: Harcourt, Brace & World, 1938.

[116] Mumford, L. *The City in History: Its Origins, Its Transformations, and Its Prospects*[M]. New York: Harcourt, Brace & World, 1961. 刘易斯・芒福德 . 城市发展史：起源、演变和前景 [M]. 宋俊岭，倪文彦，译 . 北京：中国建筑出版社，2004

[117] Niemela, J. Ecology and urban planning[J]. *Biodiversity and Conservation*, 1999, (8): 119–131.

[118] Olmsted, F. L. *Public Parks and the Enlargement of Towns*[M]. American Social Science Association, 1870.

[119] O' Sullivan, A. *Urban Economics*[M]. 5th edition. New York: McGraw–Hill, 2002.

[120] Pacione, M. *Urban Geography*[M]. London and New York: Routledge, 2001.

[121] Perry, C. *The Neighborhood Unit. Neighborhood and Community Planning: Regional Survey Volume Ⅶ, Regional Plan of New York and Its Environs*[M]. New York: Russel Sage Foundation, 1929.

[122] Popper, K. R. *Conjectures and Refutations: The Growth of Scientific Knowledge*[M]. London: Routledge & Kegan Paul, 1963.

[123] Register, R. *Eco–city Berkeley: Building Cities for a Healthier Future*[M]. CA: North Atlantic Books, 1987.

[124] Reiner, T. A. *The Place of the Ideal Community in Urban Planning*[M]. Philadelphia: University of Pennsylvania Press, 1963.

[125] Roseland, M. Dimensions of the eco–city[J]. *Cities*, 1997, 14(4): 197–202.

[126] Sager, T. *Communicative Planning Theory*[M]. Aldershot: Avebury, 1994.

[127] Savage, M., A. Warde, and K. Ward. *Urban Sociology, Capitalism and Modernity*[M]. 2nd Edition. Palgrave: Macmilan.

[128] Sitte, C. 1889. *City Planning According to Artistic Principles*[M]. Westport: Hyperion Press, 1979.

[129] Stalley, M. *Patrick Geddes: Spokesman for Man and the Environment*[M]. New Brunswick, NJ: Rutgers University Press, 1972.

[130] Sutcliffe, A. *Towards the Planned City*[M]. Oxford: Blackwell, 1981.

[131] Taylor, N. *Urban Planning Theory since 1945*[M]. Thousand Oaks: Sage, 1998. 尼格尔・泰勒 . 1945 年后西方城市规划理论的流变 [M]. 李白玉，陈贞 , 译 . 北京：中国建筑工业出版社，2006.

[132] Tetlow, J. and A. Goss. *Homes, Towns and Traffic*[M]. 2nd edition. London: Faber, 1968.

[133] Unwin, R. 1909. *Town Planning in Practice: An Introduction to the Art of Designing Cities and Suburbs*[M]. Princeton: Princeton University Press, 1993.

[134] Venturi, R. *Complexity and Contradiction in Architecture*[M]. New York: Museum of Modern Art, 1966 (2nd edition 1977).

[135] Wheeler, S. and T. Beatley. *The Sustainable Urban Development Reader*[M]. London and New York: Routledge, 2004.

[136] Wright, F. L. *Broadacre City: A New Community Plan*[M]. Scottsdale: Frank Lloyd Wright Foundation, 1935.

后记

本书开始酝酿于2018年1月，今天与读者见面，历经4年有余；如果从我2011年1月开始GRE词汇研究算起，则有12年多的时光。十年磨一剑，相信本书会对同学们有所帮助。

本书虽名为《城乡规划专业英语》，然其上篇（“核心词汇分类”）却具有普适性，即适合于各专业的英语学术文献的阅读。近几年，基于这个词表，我给国内外不少同学（不同专业背景）讲过GRE词汇、填空与阅读，广受好评，即为例证。因此，本书不仅可以作为城乡规划、人文地理本科生、研究生的教材，可供建筑学、风景园林本科生、研究生参考，而且可以作为各专业本科生、研究生学习英语的工具书。

于洪蕾、田华、朱一荣三位老师为城乡规划专家，在城乡规划学科领域具有精深的造诣，他们的思想、提供的资料深刻地影响了我，很大程度上决定了本书下篇（“经典文献导读”）的结构。感谢他们的贡献！

感谢中国建筑工业出版社的厚爱！其意见、建议与编辑使得本书的质量得到提升！

阎传海

青岛理工大学建筑与城乡规划学院

2022年3月5日